Experimental Fluid Mechanics

R. J. Adrian · M. Gharib · W. Merzkirch
D. Rockwell · J. H. Whitelaw

Springer

Berlin
Heidelberg
New York
Barcelona
Budapest
Hong Kong
London
Milan
Paris
Singapore
Tokyo

Nikita A. Fomin

Speckle Photography for Fluid Mechanics Measurements

With 83 Figures

Springer

Author

Prof. Nikita A. Fomin
Belarus Academy of Sciences
Heat and Mass Transfer Institute
P. Brovka Street 15
220072 Minsk
Belarus, CIS
e-mail: fomin@hmti.ac.by

ISBN 3-540-63767-2 Springer-Verlag Berlin Heidelberg New York

Cataloging-in-Publication Data applied for
Die Deutsche Bibliothek - CIP-Einheitsaufnahme
Fomin, Nikita A.:
Speckle photography for fluid mechanics measurements /Nikita A. Fomin. -
Berlin; Heidelberg; New York; Barcelona; Budapest; Hong Kong; London; Milan; Paris;
Singapore; Tokyo: Springer, 1998
(experimental fluid mechanics)
ISBN 3-540-63767-2

Typesetting: Camera-ready by author
Cover design: Design & Production, Heidelberg
SPIN:10552059 61/3020-5 4 3 2 1 0 - Printed on acid -free paper

Series Editors

Preface

Starting from Burch and Tokarskii's experiments in 1968 [1], during the last three decades *speckle phenomena* have been extensively investigated for use in *speckle metrology*. The results of the first decade of these investigations were summarized in the first book on the subject which appeared in 1978 (R.K.Erf, ed. *Speckle Metrology*)[2].

During the second decade the field grew and was subdivided into several new metrology areas, like *particle image velocimetry* (PIV), *dynamic speckle methods*, *speckle photography*, and *speckle tomography*. The advances in CCD imaging and computer hardware stimulated the development of *electronic speckle interferometry* (ESPI) and *digital* PIV (DPIV). Holographic PIV (HPIV) and holographic speckle photography (HSP) were introduced recently into this area of metrology.

The application areas of this metrology were also essentially enlarged. Starting from surface inspection, surface displacement, vibration and strain measurements, speckle-based techniques became well established methods in solid and fluid mechanics, astronomy, informatics, and bio-mechanics. Many books have already been published in this field [3-6]. There are a number of fundamental books in very closely related fields of optical diagnostics in fluid flows [7-13]. This book is devoted to the application of speckle photography for fluid mechanical measurements.

Speckle photography nowadays is widely used for fluid mechanics measurements. As in the other applications, the work which gave rise to employing speckle photography in metrology, was the publication of the results of observing diffraction halo in two speckle-fields, registered on the same photographic plate and shifted mutually by a short distance equal to several speckle sizes [1]. The pattern of this diffraction appeared to be identical to the diffraction pattern in two orifices, and it represents the diffraction halo, modulated by Young's interferometric fringes. The distance between these fringes and the angle of their inclination with respect to a coordinate axis are determined uniquely by the shift vector of the speckle field on the photographic plate, which enables, by analysing the diffraction halo, the establishment of the magnitude of the shift with the direction remaining ambiguous. Nowadays, two different optical principles, both of which are based on the speckle technique, are used for fluid mechanical measurements.

One of the methods, *speckle velocimetry*, relies on the use of tracer particles whose motion in the plane of the flowfield is recorded on a photographic plate. To perform these measurements, the laser-light sheet is passed through the flow section, and the seed particles are photographed perpendicular to the plane of the light sheet. Thus, the velocity vectors are determined at each point in the light sheet by a point-by-point interrogation of the obtained photograph.

The second method, *speckle photography*, is based on measuring the deflection angles of the light transmitted through the flow. This technique is a *line-of-sight* method and provides information integrated along the optical path.

The first descriptions of this method were given by Köpf [14,15], by Debrus *et al.* [16], and by Mallick and Roblin [17] as early as 1972, but it was more than a decade before applications of the new technique to fluid mechanical problems were reported [18]. The state of the art of this technique is reviewed in recent publications [19,20].

It is essential that both PIV and line-of-sight speckle photography provide quantitative information for turbulent flows as well. This is due to the universal pattern of Young's interferometric fringes which remain straight and equidistant even for turbulent flows. Such interferometric patterns can be easily read into computer memory using CCD cameras. Direct optical data input into computer memory opens the way to the storage of large amounts of experimental information. This allows one to use these data in tomographic reconstruction of 3D temperature fields and to perform statistical analysis of the flowfield providing data about turbulence.

The purpose of this book is to discuss the modern possibilities of speckle photography based on automated computer-aided specklegram processing, both for the reconstruction of complex 3D refraction index fields and for the estimation of turbulence parameters in fluid mechanical measurements.

The book is divided into 10 chapters. The first six contain a description of speckle photography and the last four contain a description of the application of speckle photography to fluid mechanical measurements. According to this the first chapter (Introduction) contains a very short description of speckle photography, the second one is devoted to diffraction theory, and the third one to speckle statistics. This is a theoretical introduction to speckle photography. The next three chapters are introductions to the experimental technique. The fourth chapter contains the description of speckle generators, light sources, and recording methods for speckle photography. The fifth one contains a description of basic optical schemes for obtaining specklegrams, whereas the sixth one deals with the automated specklegram treatment. The last four chapters are devoted to a discussion of density (temperature) measurements in laminar flows, velocity measurements, and estimation of turbulence parameters in turbulent flows. The last chapter contains a description of speckle tomography, the most promising approach in the tomography of complex phase objects.

The original idea of writing this book was a result of an early attempt to translate into the English language the author's book entitled *Speckle Photography For Fluids Flows Measurements* and published by Nauka i Technika Press, Minsk in 1989 (in Russian) [21]. However, during the decade that has passed since that time, a great number of new results in the field have been obtained. Major progress was observed in the measurement of turbulence parameters and in speckle tomography. These results and approaches occupy a significant part of this book.

For a long time the author's team has had close scientific contacts with Professor Wolfgang Merzkirch's team at the University of Essen. Many measurements were made jointly and many joint projects were developed during the last 15 years. The author is very proud of these relations and would like to express to Wolfgang Merzkirch his deepest gratitude for the friendship and scientific leadership in this very interesting 'speckle field'.

A number of grants and joint projects were very important for the development of the speckle photography technique at the Heat and Mass Transfer Institute, Minsk, Belarus. One of the first is the grant of the International Science Foundation (Soros Fond) RWB 000 as well as the subsequent grant of the Soros Fond and Belorussian Government (RWB 300). I am deeply grateful to the INTAS Foundation for grant INTAS 93-0344 and personally to Professor Clive Greated, Co-ordinator of this project. A number of publications were prepared under the scope of this project and scientific ties have been established between the Heat and Mass Transfer Institute and the University of Edinburgh. During the last decade, joint research projects on speckle photography were developed also with the University of Poitiers under guidance of Professor F.Fisson and Professor J.-B. Saulnier. Last but not least is NATO grant HTECH.LG 961001, scientific co-ordinator Professor W.Merzkirch, which allowed us to extend contacts with Essen University.

I am happy to express my appreciation and thanks to all my colleagues and friends who encouraged and supported me their help in preparing this manuscript, and especially to Prof. P. Arroyo, Drs. C. Gray, G. Blinkov, D. Vitkin, U. Wernekinck, and E. Lavinskaya.

Nikita A. Fomin
Heat and Mass Transfer Institute
Minsk, Belarus

November 1997

Table of Contents

1. Introduction

Modern speckle photography based on the computer-aided evaluation of speckle patterns extends the methods of optical flow visualization and diagnostics, and it allows the quantitative derivation of a two dimensional map of the angles of deflection experienced by light when passing through a flow under study. Such a map may be used for the tomographic reconstruction of 3D density (or temperature) fields in both laminar and turbulent flows.

Two prominent features of speckle photography will be elucidated in the Introduction. The first one is the availability of high data densities due to the possibility of generating very small scale speckles. The typical speckle size is comparable to the wavelength of the radiation forming the speckle pattern, and this means that 1 mm^2 of a *specklegram* (the photograph of a speckle pattern) can include more than a million of individual speckles. With the *point-by-point* analysis of the specklegram, performed for a large amount of interrogation spots, one may obtain an instantaneous, two-dimensional map of the flow with a very high spatial resolution. Further below we shall demonstrate this capability in conjunction with the discussion of two possible optical schemes for the formation of speckles.

The second prominent feature provided by speckle photography is the possibility of easily transferring into memory of a computer the quantitative information obtainable from the patterns of Young's interference fringes when these patterns are observed with a standard CCD camera. Young's fringes modulate the diffraction zone ('halo') behind a double exposure specklegram that is interrogated, point-by-point, with the thin beam of a CW laser for the purpose of evaluation. With this procedure it is required to determine the orientation (angle) and the spacing of the fringes; this can be done easily with image processing methods, because the fringes form a universal pattern, equally-spaced and parallel. A great amount of experimental information can be stored in this way; e.g., when recording a specklegram on a standard holographic plate and interrogating the specklegram with the laser beam in steps of 0.1 mm, one may store the values of more than 10^6 vectors. Such a vectors may consist of the two components of the deflection angle of the light transmitted through the flow. In this example the recorded and stored data could be used for a tomographic reconstruction of the 3D density (temperature) distribution in the investigated flow, and a statistical analysis of this data might provide information on the turbulence characteristics of this flow.

The two mentioned features of speckle photography will now be considered in more detail.

1.1 Two schemes of speckle formation

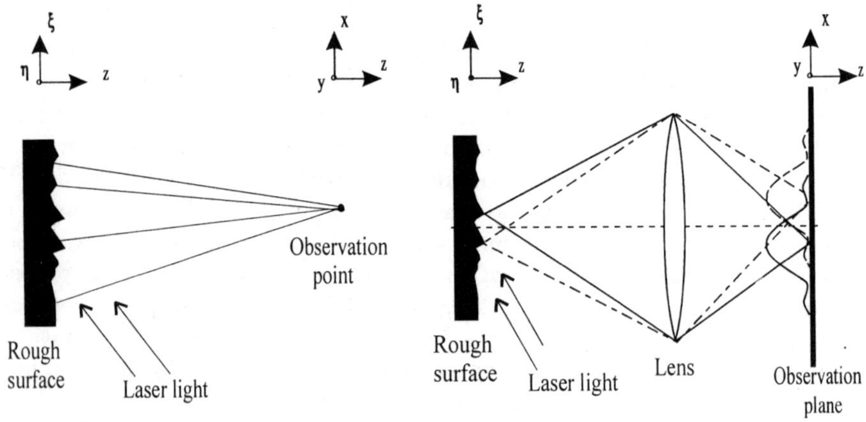

Fig. 1.1. Two schemes of speckle pattern formation under illumination of a rough surface by laser light

A speckle pattern is formed either by the reflection of coherent light from a rough surface, or by transmitting the light through a transparent medium having a randomly fluctuating refractive index distribution and by subsequent 3D multiple beam interference. The speckle pattern is characterized by a random intensity distribution that may be described by statistical means [22]. The speckles are clearly visible when such an object is viewed under illumination by laser light.

Fig. 1.1 illustrates two possible ways of speckle formation when laser light is reflected from a rough surface.

The simplest description of speckle formation may be made using the set of assumptions summarised by Dainty [5]. The assumptions are

- the light is perfectly coherent,
- the rough surface introduces phase fluctuations greater than 2π,
- the surface does not depolarize the light,
- a large number of scattering centres contribute to the intensity at the point in the observation plane.

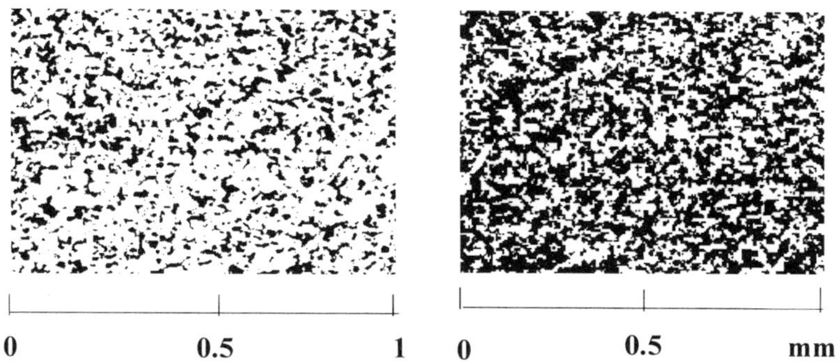

Fig. 1.2. Small fragments of the single exposure (on the left) and double exposure (on the right) specklegrams under high electro-optical magnification

For free space propagation (Fig. 1.1, on the left), the optical wave resulting at any moderately distant point consists of many coherent contributions or wavelets, each arising from a different microscopic element of the surface [22]. 3D interference of these wavelets results in the granular pattern of intensity that are called a speckle pattern. Small fragments of the speckle patterns under high magnification are shown in Fig. 1.2.

The speckle pattern formed for a free space geometry is called *objective speckles.* The size of the speckles, a statistical average of the distance between adjacent regions of maximum and minimum intensity, for this method of speckle pattern formation in first approximation is [23]

$$\sigma_{\mathrm{sp}} = 1.22\lambda\frac{z}{D} \tag{1.1}$$

where λ is the wavelength of the radiation forming a speckle pattern, z is the distance from the rough surface to the observation point and D is diameter of the illuminated part of the rough surface.

If, alternatively, the speckle pattern is formed by collecting the scattered radiation with a lens and focusing it on to the screen (Fig. 1.1b), a *subjective* speckle pattern is formed. The size of the averaged speckles is then related to the f/number of the lens, also called the aperture ratio F of the lens, and to the magnification M at which the lens is operating [23]:

$$\sigma_{\mathrm{sp}} = 1.22\lambda(1+M)F = 1.22\lambda(1+M)\frac{f}{D} \tag{1.2}$$

with f being the focal length of the lens and D its diameter.

The subjective speckle pattern may be treated as the image of a speckle pattern of different scale in the plane of the scattering surface. For these speckles the averaged size is given by

$$\sigma'_{\rm sp} = 1.22\lambda \frac{(1+M)}{M} F = 1.22\lambda \frac{(1+M)}{M} \frac{f}{D}. \tag{1.3}$$

So, for both schemes of speckle formation it can be seen that it is possible to obtain a speckle size comparable with the wavelength of the radiation forming the speckle pattern.

1.2 Original way of specklegram treatment

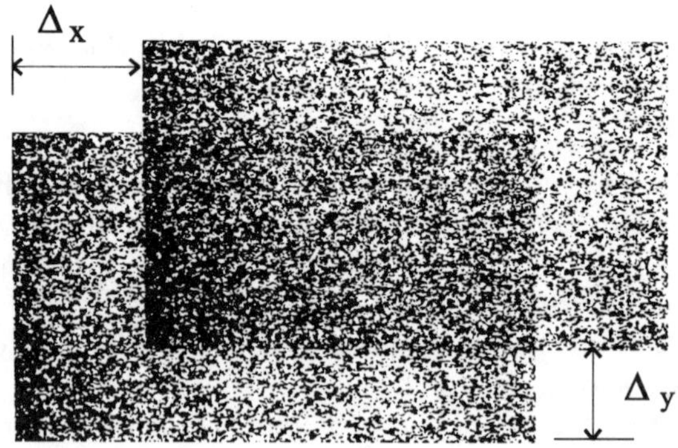

Fig. 1.3. A model of a double exposure specklegram.

Taking in turn a sequence of instantaneous images of the moving speckle pattern on a photograph, called *a multi-exposure specklegram*, it is possible to determine the average speckle displacement at each small area of the specklegram. The motion of the speckles can be related to the movement or deformation of a rough surface, which generates the speckle pattern. Thus, the movement or deformation of a rough surface can be quantitatively measured by illumination of this surface by coherent light and recording of the movement of the speckle pattern.

The starting point of utilizing this technique in metrological application was Burch and Tokarski's publication in 1968 [1]. They have shown that the diffraction halo of any small part of a *double exposure specklegram* is the same as the diffraction halo in Young's interferometer with a screen having two orifices, which diameter is equal to an average speckle size and the spacing between which is equal to the speckle shift at the treated point of the specklegram.

Let us suppose that two exposures of the speckle pattern on the photographic plate were made and between these exposures the speckles were shifted, for instance, by the distance $\mathbf{\Delta s} = \mathbf{i}\cdot\Delta_x + \mathbf{j}\cdot\Delta_y$, as shown in Fig. 1.3.

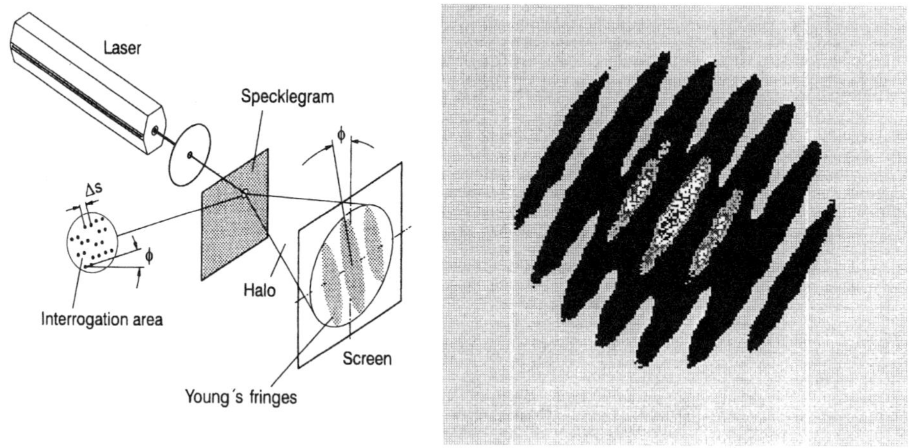

Fig. 1.4. The formation of Young's fringes under illumination of a small fragment of a double-exposure specklegram (on the left) and an example of Young's fringes in quasi-colour presentation (after primary computer-aided smoothing)

As we will see in the next chapter, the intensity distribution in the diffraction halo under the illumination of a small fragment of the double exposure specklegram by a thin CW laser beam is Young's interferometric fringes (see Fig. 1.4). This distribution is

$$I(\mathbf{r}) = I_{\mathrm{h}}(\mathbf{r}) \cdot \cos^2\left[\frac{k(\mathbf{r} \cdot \boldsymbol{\Delta}\mathbf{s})}{2L}\right] \tag{1.4}$$

where $\mathbf{r}$ is the coordinate in the observation plane: $\mathbf{r} = \mathbf{i} \cdot x + \mathbf{j} \cdot y$, k is the *wavenumber* ($k = \frac{2\pi}{\lambda}$), L is the distance from the specklegram to the observation screen, and $I_h(\mathbf{r})$ is the diffraction halo of a *single exposure specklegram.* It is seen that the diffraction halo of a single exposure specklegram is now modulated by a series of uniformly spaced cosine square fringes whose spacing is inversely proportional to $|\boldsymbol{\Delta}\mathbf{s}|$, and whose direction is normal to $\boldsymbol{\Delta}\mathbf{s}$.

The dark fringes are observed when

$$\mathbf{r} \cdot \boldsymbol{\Delta}\mathbf{s} = (n + \frac{1}{2})\lambda L, \quad n = 0, \pm 1, \pm 2, \ldots \tag{1.5}$$

and bright fringes when

$$\mathbf{r} \cdot \boldsymbol{\Delta}\mathbf{s} = n\lambda L, \quad n = 0, \pm 1, \pm 2, \ldots \tag{1.6}$$

Eqs. 1.5 and 1.6 can be combined as

$$|\boldsymbol{\Delta}\mathbf{s}| = \frac{\lambda L}{d_{\mathrm{f}}}, \tag{1.7}$$

where d_{f} is the fringes spacing.

Thus, using the original way of the specklegram treatment, it is possible, with very high precision, to determine at each interrogation point of the specklegram the vector of the speckle shift.

2. Diffraction Theory of Speckle Photography

The effects of diffraction are very important both for speckle formation during specklegram recording and for optical transformation of the wavefront during specklegram processing. Moreover, the statistical description of the intensity distribution in a speckle pattern is based of the diffraction analysis. So, in order to offer a brief introduction to speckle photography, we will start from the general principles of diffraction theory following mainly the notations of Born and Wolf [24] and Goodman [25].

2.1 Maxwell's equations

To outline the diffraction of light in non-homogeneous media, one starts with *Maxwell's equations.* These equations describe the propagation of any disturbances of the electromagnetic field both in a vacuum and in dielectric and in the absence of volume charges and currents (free space propagation) are

$$\mathrm{rot}\mathbf{H}(\mathbf{r},t) - \frac{1}{c}\frac{\partial\mathbf{D}(\mathbf{r},t)}{\partial t} = 0 \tag{2.1}$$

$$\mathrm{rot}\mathbf{E}(\mathbf{r},t) + \frac{1}{c}\frac{\partial\mathbf{B}(\mathbf{r},t)}{\partial t} = 0 \tag{2.2}$$

$$\mathrm{div}\mathbf{D}(\mathbf{r},t) = 0 \tag{2.3}$$

$$\mathrm{div}\mathbf{B}(\mathbf{r},t) = 0. \tag{2.4}$$

The electromagnetic field is determined by the electric vector $\mathbf{E}$, the electric displacement $\mathbf{D}$, the magnetic vector $\mathbf{H}$, and the magnetic induction $\mathbf{B}$.

Maxwell's equations are supplemented by the 'material' equations which describe the behavior of substances under the influence of the field. For *isotropic* materials these equations are

$$\mathbf{D} = \varepsilon\mathbf{E} \tag{2.5}$$

$$\mathbf{B} = \mu\mathbf{H}, \tag{2.6}$$

where ε is the dielectric constant (or permittivity), and μ is the magnetic permeability.

Maxwell's equations can be transformed in to:

$$\bigtriangledown^2\mathbf{E} - \frac{\varepsilon\mu}{c^2}\ddot{\mathbf{E}} + [\bigtriangledown(\log\mu)] \times \mathrm{rot}\mathbf{E} + \bigtriangledown[\mathbf{E} \cdot \bigtriangledown(\log\varepsilon)] = 0 \tag{2.7}$$

$$\bigtriangledown^2\mathbf{H} - \frac{\varepsilon\mu}{c^2}\ddot{\mathbf{H}} + [\bigtriangledown(\log\varepsilon)] \times \mathrm{rot}\mathbf{H} + \bigtriangledown[\mathbf{H} \cdot \bigtriangledown(\log\mu)] = 0. \tag{2.8}$$

Here the operators $\bigtriangledown$ and $\bigtriangledown^2$ are vectors with components $(\frac{\partial}{\partial x}, \frac{\partial}{\partial y}, \frac{\partial}{\partial z})$ and $(\frac{\partial^2}{\partial x^2}, \frac{\partial^2}{\partial y^2}, \frac{\partial^2}{\partial z^2})$, respectively.

For a homogeneous medium these equations are simplified to:

$$\bigtriangledown^2\mathbf{E} - \frac{\varepsilon\mu}{c^2}\ddot{\mathbf{E}} = 0, \tag{2.9}$$

$$\bigtriangledown^2\mathbf{H} - \frac{\varepsilon\mu}{c^2}\ddot{\mathbf{H}} = 0, \tag{2.10}$$

which describes an electromagnetic wave propagating with a velocity v:

$$v = \frac{c}{\sqrt{\varepsilon\mu}} = \frac{c}{n}. \tag{2.11}$$

Here n is the refractive index (or the index of refraction).

This simplification can be done also for 'smooth' variations of the index of refraction, and, as shown by Clifford [26], is valid even for spatial scales of refractive index variations comparable with the wavelength of the radiation.

A wave can be represented as a sum of harmonic waves, for which it is convenient to use complex variables:

$$\mathbf{V}(\mathbf{r}, t) = \mathrm{Re}\ \{\mathbf{U}(\mathbf{r}) \cdot \exp(-\mathrm{i}\omega t)\}, \tag{2.12}$$

where $\mathbf{V}$ is one of the $\mathbf{E}$ or $\mathbf{H}$ vectors and $\mathbf{U}(\mathbf{r})$ is *the complex amplitude* of the wave. The latter value can be written as:

$$\mathbf{U}(\mathbf{r}) = \mathbf{U}(x, y, z) = U(x, y, z) \cdot \exp[\mathrm{i}\Theta(x, y, z)] \tag{2.13}$$

with $U(x, y, z)$ and $\Theta(x, y, z)$ being the amplitude and phase of the wave, respectively.

Using the following relations one can determine the optical frequency, ν, the wavelength, λ, and the wavenumber, k, of the wave:

$$\nu = \frac{\omega}{2\pi} = \frac{1}{T}; \ \lambda = v \cdot \frac{2\pi}{\omega} = v \cdot T; \ k = \frac{2\pi}{\lambda} = \frac{\omega}{v} = \frac{n\omega}{c} \tag{2.14}$$

It is also convenient to use the vector wavenumber $\mathbf{k}$ whose direction corresponds to that of wave propagation and whose value is equal to k. For the complex amplitude, $\mathbf{U}$, we have the equation:

$$\nabla^2 \mathbf{U} + k^2 \mathbf{U} = 0. \tag{2.15}$$

Equation (2.15) is also known as the Helmholtz equation. For a monochromatic wave this equation can be written in scalar form

$$\nabla^2 U + k^2 U = 0. \tag{2.16}$$

This equation will be used in the next section for obtaining Kirchhoff's integral and later (see chapter 10) for describing laser light propagation through turbulence, see e.g. [26-28].

Assuming a geometrical optics approximation ($\lambda \to 0$) the energy may be regarded as being transported along certain curves (rays).

Maxwell's equations for this case can be transformed into the eikonal equation

$$(\frac{\partial \mathcal{E}}{\partial x})^2 + (\frac{\partial \mathcal{E}}{\partial y})^2 + (\frac{\partial \mathcal{E}}{\partial z})^2 = n^2(x, y, z) \tag{2.17}$$

with $\mathcal{E}$ being an eikonal, a real scalar function of position. This function is simply connected with the wave phase, see Eq.(2.13):

$$\mathcal{E}(x, y, z) = \frac{\Theta(x, y, z)}{k_0}, \tag{2.18}$$

where k_0 is the wavenumber in vacuum.

The surfaces $\mathcal{E}(x, y, z) =$ constant may be called the geometrical wave surfaces or the geometrical wavefronts. The eikonal equation was derived by using the first-order Maxwell's equations, but it may also be derived from the second-order wave equations for the electric or magnetic field vectors [24].

It is convenient to introduce a unit vector $\mathbf{s}$, as

$$\mathbf{s} = \frac{1}{n} \nabla \mathcal{E}. \tag{2.19}$$

The geometrical light rays may now be defined as the orthogonal trajectories to the geometrical wavefronts $\mathcal{E}(x, y, z) =$ constant.

Denoting by $\mathbf{r}(s)$ the radius-vector of the point situated on the light ray, the ray equation can be written in the form

$$n \frac{\mathrm{d}\mathbf{r}}{\mathrm{d}s} = \nabla \mathcal{E}. \tag{2.20}$$

The latter equation can be transformed into

$$\frac{\mathrm{d}}{\mathrm{d}s} n \frac{\mathrm{d}\mathbf{r}}{\mathrm{d}s} = \nabla n. \tag{2.21}$$

In chapter 9 this equation will be used for ray tracing through a turbulent field.

2.2 Fresnel and Fraunhofer diffraction

The distribution of the light intensity in the diffraction region after the interaction of the plane wavefront with some obstacle in the general case can be calculated using the solution of Maxwell's equations with appropriate boundary conditions. Such calculations are performed using high-power computers, see e.g. [29].

A simpler calculation under the condition when the characteristic scales of length under consideration are essentially greater than the radiation wavelength is based on the Huygens–Fresnel principle. In this case the diffraction field is determined as a superposition of elementary secondary wavelets which originate from the object under study.

Suppose that a wavelet emitted from a point $Q(\xi, \eta)$ on the (ξ, η) plane is described by function $K(\xi, \eta, 0; x, y, z)$. This function is also referred to as *kernel*.

For free space propagation the kernel has the form of a spherical wave [30]

$$K \sim \frac{exp(\mathrm{i}kR)}{R} \tag{2.22}$$

and the wave distribution on the observation plane (x, y) can be expressed for a given set of boundary conditions on (ξ, η) plane

$$U(x, y, z) = \mathrm{Const} \cdot \int\int_{(S)} K \cdot U(\xi, \eta, 0)\, \mathrm{d}S \tag{2.23}$$

In linear systems theory, K is referred to also as the impulse response of the system, see [25].

The exact representation of the kernel may be found as the solution of the Helmholtz equation (2.15). In this way, using Green's theorem Kirchhoff connects the complex amplitude of a wave at an arbitrary point $P(x, y, z)$ with the integral over a surface about this point:

$$U_P = \frac{1}{4\pi} \int\int_S \left\{ U_s \cdot \frac{\partial}{\partial n}(\frac{\exp(\mathrm{i}kR)}{R}) - \frac{\exp(\mathrm{i}kR)}{R} \frac{\partial U_s}{\partial n} \right\} \mathrm{d}S. \tag{2.24}$$

Here R is the distance from the surface element to the point P. In the case, when $k \to 0$, Kirchhoff's integral takes the form

$$U_P = \frac{1}{4\pi} \int\int_S \left\{ U_s \cdot \frac{\partial}{\partial n}(\frac{1}{R}) - \frac{1}{R} \frac{\partial U_s}{\partial n} \right\} \mathrm{d}S \tag{2.25}$$

A commonly used configuration in optical diffraction problems is shown in Fig. 2.1. An object, for instance, a screen with an aperture, is located in the (ξ, η) plane and is illuminated with a monochromatic plane wave of normal

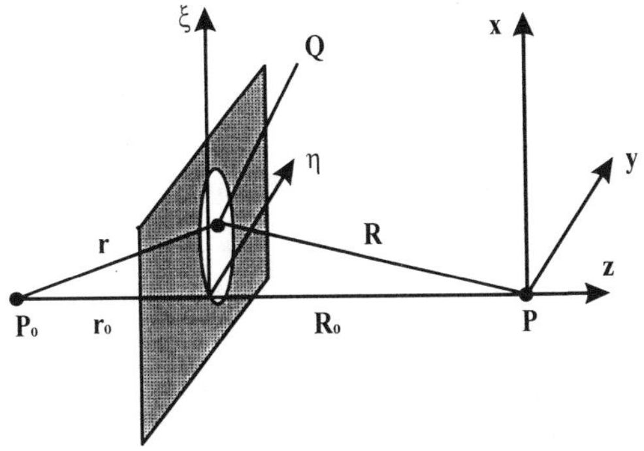

Fig. 2.1. Geometry for Kirchhoff's diffraction theory

incidence. The diffraction on the screen can be described by the Fresnel–Kirchhoff formula obtained using integral (2.25)

$$U_P = -\frac{\mathrm{i}U_{P_0}}{2\lambda}\int\int_S \frac{\exp(ik[r+R])}{rR}[\cos(n,r) - \cos(n,R)]\,dS \qquad (2.26)$$

The integration is performed over the area of the orifice, for details see, e.g. [24].

Let us now consider the situation when the distribution of the wave amplitude is known on some plane (ξ, η). For simplicity, assume $U(\xi,\eta) = U_S$ within the aperture (see Fig. 2.1) and $U(\xi,\eta) = 0$ elsewhere on the plane.

At the point $P(x,y)$ on the plane (x,y), spaced at a distance $z = R_0$ from the screen, the field amplitude will be given by

$$U(x,y) = \frac{\mathrm{i}}{\lambda}\cdot\int_{-\infty}^{\infty}\int_{-\infty}^{\infty} U_S \frac{\exp[-\mathrm{i}kR]}{R}\mathrm{d}\xi\mathrm{d}\eta, \qquad (2.27)$$

where

$$R = \sqrt{R_0^2 + (x - \xi)^2 + (y - \eta)^2}.$$

Formula (2.27) express quantitatively Huygens' principle of superposition.

The diffraction effects are significant in the cases when aperture diameters are comparable with the wavelength of the radiation, and, therefore, in most cases of practical interest $R \simeq R_0$. The following approximation is based on this fact and can be applied when $x \ll z$ and $y \ll z$.

The Fresnel approximation corresponds to the case when

$$R \approx R_0 \cdot \left[1 + \frac{(x-\xi)^2}{2R_0^2} + \frac{(y-\eta)^2}{2R_0^2}\right]. \qquad (2.28)$$

By substituting (2.28) into (2.27) we obtain

$$U(x,y) = \frac{\mathrm{i}\cdot\exp(\mathrm{i}kR_0)}{\lambda R_0}\cdot\int_{-\infty}^{\infty}\int_{-\infty}^{\infty} U_S \times \exp -\mathrm{i}k\left[\frac{(x-\xi)^2+(y-\eta)^2}{2R_0}\right]\mathrm{d}\xi\mathrm{d}\eta. \tag{2.29}$$

It is possible to show that this approximation will be valid starting from the distance [1]

$$R_0^3 \gg \frac{(x-\xi)^4+(y-\eta)^4}{\lambda}. \tag{2.30}$$

For the wavelength $\lambda = 1\,\mu m$ and diameter of aperture 1 mm, the Fresnel approximation will describe the diffraction at a distance of about $1cm$ from the screen.

In the far field, when, for instance,

$$k\cdot\frac{(\xi^2+\eta^2)\mathrm{max}}{2z} \ll \pi, \tag{2.31}$$

the quadratic terms in (2.28) can be omitted and the integral (2.29) can be rewritten in the form

$$U(x,y,z) = \frac{\mathrm{i}\cdot\exp(\mathrm{i}kz)}{\lambda z}\cdot\exp(-\mathrm{i}k\frac{x^2+y^2}{2z})\cdot\int_{-\infty}^{\infty}\int_{-\infty}^{\infty} U_S \times \exp -\mathrm{i}k\left[\frac{x\xi+y\eta}{z}\right]\mathrm{d}\xi\mathrm{d}\eta. \tag{2.32}$$

This approximation is called the Fraunhofer approximation and is valid in the far field, where

$$z \gg \frac{D^2}{\lambda}, \tag{2.33}$$

and D is the aperture diameter. Notice that condition (2.33) is indeed more stringent than (2.30).

As an example of Fresnel diffraction we can consider the distribution of the light intensity behind the screen with a sharp edge.

Let us consider now that the amplitude distribution on the screen is described by the equation

$$U(\xi,\eta) = U_0\cdot\frac{\exp(-\mathrm{i}kr_0)}{r_0}, \tag{2.34}$$

i.e. we have a point light source separated from the screen by the distance r_0.

Using Eq. (2.26) for this case we obtain

[1] Equation (2.30) gives a sufficient condition for the Fresnel approximation, see [31,32].

$$\begin{aligned} U(x,y) &= U_1 \times \int_{-\infty}^{\infty} \exp\left\{-ik\left[\frac{(x-\xi)^2}{2R_0}+\frac{\xi^2}{2r_0}\right]\right\} d\xi \\ &\quad \times \int_{-\infty}^{\infty} \exp\left\{-ik\left[\frac{(y-\eta)^2}{2R_0}+\frac{\eta^2}{2r_0}\right]\right\} d\eta, \end{aligned} \tag{2.35}$$

where

$$U_1 = U(\xi,\eta)\cdot\frac{\mathrm{i}\cdot\exp(-\mathrm{i}kR_0)}{\lambda R_0}.$$

The dependence of the wave amplitude on the coordinate x is given by

$$U(x) = U_2\cdot\int_0^{\infty} \exp\left\{-ik\left[\frac{(x-\xi)^2}{2R_0}+\frac{\xi^2}{2r_0}\right]\right\} d\xi, \tag{2.36}$$

where

$$U_2 = \mathrm{Const}(x).$$

For a laser source or for a point light source at infinity, $r_0 \to \infty$, and, therefore,

$$U(x,y) = U_2\cdot\int_0^{\infty} \exp\left[-ik\frac{(x-\xi)^2}{2R_0}\right] d\xi. \tag{2.37}$$

Introducing a new variable, u,

$$u = \sqrt{\frac{2}{\lambda R_0}}\cdot(x-\xi),$$

we rewrite expression (2.37) in the form

$$\begin{aligned} U(x) &= A_3\cdot\left[\int_{-\infty}^{0}\exp\left(-\mathrm{i}\tfrac{\pi}{2}u^2\right)du + \int_0^{w}\exp\left(-\mathrm{i}\tfrac{\pi}{2}u^2\right)du\right] \\ &= U_3\cdot\left[\tfrac{1}{2}(1-\mathrm{i}) + \mathcal{C}(w) - \mathrm{i}\mathcal{S}(w)\right]. \end{aligned} \tag{2.38}$$

The intensity distribution in the diffraction zone is then defined by

$$I(x) = |U(x)|^2 = U_4\cdot\left\{\left[\mathcal{C}(w)+\frac{1}{2}\right]^2 + \left[\mathcal{S}(w)+\frac{1}{2}\right]^2\right\}. \tag{2.39}$$

Here $U_3 = \mathrm{Const}(x), U_4 = \mathrm{Const}(x), w = x\cdot\sqrt{2/\lambda R_0}$, and $\mathcal{C}(w), \mathcal{S}(w)$ are Fresnel's integrals:

$$\mathcal{C}(w) = \int_0^{w}\cos\frac{\pi u^2}{2}du,$$

$$\mathcal{S}(w) = \int_0^{w}\sin\frac{\pi u^2}{2}du.$$

It is easily seen that the intensity distribution far from the edge, where diffraction is negligible, is

$$I_0 = I(+\infty) = A_4\left\{\left[\mathcal{C}(\infty)+\frac{1}{2}\right]^2 + \left[\mathcal{S}(\infty)+\frac{1}{2}\right]^2\right\} = 2A_4. \tag{2.40}$$

Therefore,

$$\frac{I(x)}{I_0} = \frac{1}{2} \cdot \left\{ \left[\mathcal{C}(w) + \frac{1}{2} \right]^2 + \left[\mathcal{S}(w) + \frac{1}{2} \right]^2 \right\}. \tag{2.41}$$

As an example of Fraunhofer diffraction we will consider diffraction at a small aperture.

In the case of a rectangular aperture with sides $2a \times 2b$, the Fraunhofer diffraction diagram will be determined by

$$\begin{aligned} U(x,y,z) &= \text{Const} \cdot \int_{-a}^{a} \int_{-b}^{b} \exp[-\tfrac{\mathrm{i}k(x\xi+y\eta)}{z}]\,\mathrm{d}\xi\mathrm{d}\eta \\ &= 4 \cdot \text{Const} \cdot \tfrac{\sin(kxa/z)}{kxa/z} \tfrac{\sin(kyb/z)}{kyb/z} \end{aligned} \tag{2.42}$$

and for the intensity distribution we have

$$I(x,y,z) = I_0 \left[\frac{\sin(kxa/z)}{kxa/z} \frac{\sin(kyb/z)}{kyb/z} \right]^2. \tag{2.43}$$

In the case of a circular aperture it is convenient to use a polar coordinate systems $(\rho, \theta), (r, \psi)$ both at the aperture and at the observation plane:

$$\begin{aligned} \xi &= \rho \cdot \cos\theta, \\ \eta &= \rho \cdot \sin\theta, \\ x &= r \cdot \cos\psi, \\ y &= r \cdot \sin\psi \end{aligned}$$

and the light amplitude in the far field is given by

$$U(r,\psi) = \text{Const} \cdot \int_0^D \int_0^{2\pi} \exp[-\mathrm{i}k\rho r \cos(\theta - \psi)]\rho\,\mathrm{d}\rho\mathrm{d}\theta \tag{2.44}$$

with D being aperture diameter. Integration of Eq. (2.44) leads to

$$U(r) = \text{Const} \cdot \frac{2\mathcal{J}_1(kDr/z)}{kDr/z} \tag{2.45}$$

and for the light intensity distribution we obtain

$$I(r) = I_0 \left[\frac{2\mathcal{J}_1(kDr/z)}{kDr/z} \right]^2. \tag{2.46}$$

Here $r^2 = x^2 + y^2$ and $\mathcal{J}_1$ is the Bessel function of the first order.

These two examples of Fraunhofer diffraction are illustrated by Fig. 2.2.

Analysing light distributions for spherical waves it is convenient to use spherical coordinate system:

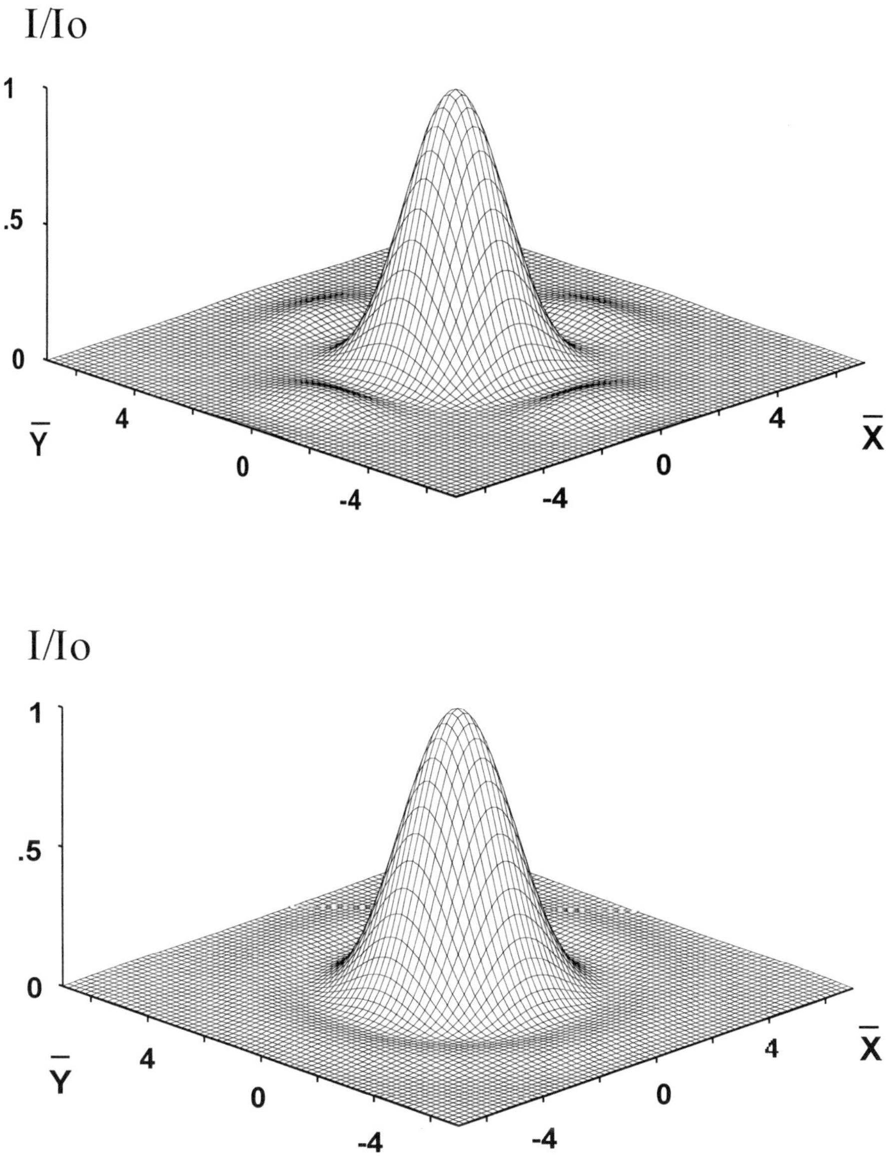

Fig. 2.2. Light intensity distribution in Fraunhofer diffraction zone for small rectangular (top) and circular (bottom) apertures

$$
\begin{aligned}
x &= R\sin\theta\cos\varphi, \\
y &= R\sin\theta\sin\varphi, \\
z &= R\cos\theta,
\end{aligned}
$$

and the light amplitude in the far field will be given by

$$U(\theta,\varphi) = \mathrm{Const}\cdot\int_{(S)} A(\xi,\eta)\exp[ik\sin\theta\cdot(\xi\cos\varphi+\eta\sin\varphi)]\,\mathrm{d}\xi\mathrm{d}\eta. \quad (2.47)$$

In the case of a circular aperture, integration of Eq. (2.47) leads to

$$U(\theta,\varphi) = \mathrm{Const}\cdot\frac{\mathcal{J}_1\left(\frac{\pi D}{\lambda}\sin\theta\right)}{\frac{\pi D}{\lambda}\sin\theta}. \quad (2.48)$$

In the case of a rectangular aperture, the diffraction diagram will be determined by

$$\frac{U(\theta,\varphi)}{U_{max}} = \frac{\sin\left(\frac{\pi a}{\lambda}\sin\theta\cos\varphi\right)}{\frac{\pi a}{\lambda}\sin\theta\cos\varphi}\cdot\frac{\sin\left(\frac{\pi b}{\lambda}\sin\theta\sin\varphi\right)}{\frac{\pi b}{\lambda}\sin\theta\sin\varphi}. \quad (2.49)$$

Let us now consider a variation of the wave phase in the aperture plane. For instance, for the plane wavefront coming into the aperture plane at some inclination angle we will have

$$U(\xi,\eta) = \exp(-ik\xi\sin\theta_0), \quad (2.50)$$

and Eq. (2.48) will be transformed in to

$$\frac{U(\theta,\varphi)}{U_{max}} = \frac{\sin\left(\frac{\pi a}{\lambda}[\sin\theta-\sin\theta_0]\cos\varphi\right)}{\frac{\pi a}{\lambda}[\sin\theta-\sin\theta_0]\cos\varphi}\cdot\frac{\sin\left(\frac{\pi b}{\lambda}\sin\theta\sin\varphi\right)}{\frac{\pi b}{\lambda}\sin\theta\sin\varphi}. \quad (2.51)$$

As we can see, the diffraction diagram is in this case approximately similar to that described by Eq. (2.48) but has been turned through an angle θ_0. The effective aperture size is changed

$$a_{\mathrm{eff}} = a\cdot\cos\theta_0. \quad (2.52)$$

Let us examine the Fraunhofer diffraction on the aperture $\mathcal{A}_2$ whose size in the direction of the axis (ξ) is μ times as large as that for the aperture $\mathcal{A}_1$. For diffraction at the aperture $\mathcal{A}_1$ we have

$$U_1(x,y) = \mathrm{Const}\cdot\int\int_{(\mathcal{A}_1)}\exp[-\frac{ik}{z}(x\xi+y\eta)]\,\mathrm{d}\xi\mathrm{d}\eta. \quad (2.53)$$

A similar expression describes the diffraction at the aperture $\mathcal{A}_2$:

$$U_2(x,y) = \text{Const} \cdot \int\int_{(\mathcal{A}_2)} \exp[-\frac{\mathrm{i}k}{z}(x\xi + y\eta)]\,\mathrm{d}\xi\mathrm{d}\eta. \tag{2.54}$$

Using change of variables in the latter expression, we obtain:

$$U_2(x,y) = \mu\text{Const} \cdot \int\int_{(\mathcal{A}_1)} \exp[-\frac{\mathrm{i}k}{z}(\mu x\xi' + y\eta')]\,\mathrm{d}\xi'\mathrm{d}\eta' = \mu U_1(\mu x, y), \tag{2.55}$$

where

$$\xi' = \frac{1}{\mu}\xi, \quad \eta' = \eta. \tag{2.56}$$

Thus, the diffraction pattern at the aperture $\mathcal{A}_2$ is compressed by μ times as compared with the diffraction at the aperture $\mathcal{A}_1$ and, accordingly, the intensity at some point of the diffraction pattern at the aperture $\mathcal{A}_2$ is μ^2 times larger than that for the aperture $\mathcal{A}_1$. The same consideration can be repeated with respect to the other coordinate axis.

2.3 Fourier transform

Equation (2.29) has the form of a convolution integral. The Fresnel diffraction is then interpreted to be the convolution of U_S and the function $T(x,y,z)$

$$T(x,y,z) = \exp[\frac{\mathrm{i}k}{2z}(x^2 + y^2)], \tag{2.57}$$

where the function $T(x,y,z)$ is called the impulse response of the system [30]. Therefore, one can write

$$U(x,y,z) = U_S(\xi,\eta) \otimes T(x,y,z), \tag{2.58}$$

where the $\otimes$ denotes a convolution.

Let us take the Fourier transform of both sides of Eq. (2.58):

$$\mathcal{F}\{U(x,y,z)\} = \mathcal{F}\{U_S(\xi,\eta)\} \cdot \mathcal{F}\{T(x,y,z)\}. \tag{2.59}$$

The function $\mathcal{T} = \mathcal{F}\{T(x,y,z)\}$ is called the transfer function of the optical system

$$\mathcal{T} = \int_{-\infty}^{\infty}\int_{-\infty}^{\infty} T(x,y,z)\exp[-i2\pi(f_x x + f_y y)]\,\mathrm{d}x\mathrm{d}y \tag{2.60}$$

where f_x and f_y denote spatial frequency variables.

Equation (2.32) shows that in the Fraunhofer approximation the distribution of the amplitude in the diffraction field can be constructed as a Fourier transform with frequencies $f_x = x/\lambda z$ and $f_y = y/\lambda z$:

$$U(x,y,z) = C \cdot \mathcal{F}\{U_S\} \tag{2.61}$$

where C is a coefficient.

In contrast to this in the Fresnel approximation the distribution of the amplitude in the diffraction field can be constructed as a Fourier transform of a function u_s, which is different from the initial amplitude distribution:

$$U(x,y,z) = C \cdot \mathcal{F}\{u_s\} \tag{2.62}$$

where

$$u_s = U_S \cdot \exp[\frac{ik}{2z}(x^2+y^2)]. \tag{2.63}$$

The difference between the functions u_s and U_S decreases when the distance z increases. The constant C here is equal to:

$$C = \frac{\exp(ikz)}{ikz} \cdot \exp[\frac{ik}{2z}(x^2+y^2)]. \tag{2.64}$$

Using the *similarity theorem*, it is possible to express the Fourier transform of the function of the modified coordinates $g(ax, by)$ in terms of the Fourier transform of the original function, $g(x,y)$:

$$\mathcal{F}\{g(ax,by)\} = \frac{1}{|ab|} G\left(\frac{f_x}{a}, \frac{f_y}{b}\right) \tag{2.65}$$

with $G(f_x, f_y)$ being the Fourier transform of $g(x,y)$:

$$G(f_x, f_y) = \mathcal{F}\{g(x,y)\}. \tag{2.66}$$

This means that the extension of coordinates in the space domain (x,y) leads to compression of coordinates in the frequency domain (f_x, f_y) and decreases the total spectral amplitude.

2.4 Diffraction at single apertures

The distributions of the light intensity in the Fraunhofer diffraction zone may also be obtained with the help of the Fourier transform described in the previous section. Let assume that a screen has an aperture and is illuminated by a point source at infinity.

For a rectangular aperture with dimensions $2a$ and $2b$ the amplitude transmittance (*the pupil function*) of the screen is

$$t(\xi,\eta) = \mathrm{rect}\left(\frac{\xi}{2a}\right) \mathrm{rect}\left(\frac{\eta}{2b}\right) \tag{2.67}$$

and the amplitude distribution of the Fraunhofer diffraction is determined by the Fourier transform of the amplitude distribution in the aperture plane:

$$\mathbf{U}(x,y) = C \cdot \mathcal{F}\{\mathbf{U}(\xi,\eta)\} \tag{2.68}$$

where

$$C = \frac{\exp(\mathrm{i}kz) \cdot \exp\left[\mathrm{i}\frac{k}{2z}(x^2+y^2)\right]}{\mathrm{i}\lambda z}$$

Taking into account that

$$\mathcal{F}\{\mathbf{U}(\xi,\eta)\} = 4ab \cdot \mathrm{sinc}(2af_x) \cdot \mathrm{sinc}(2bf_y), \tag{2.69}$$

where

$$\mathrm{sinc}(x) = \frac{\sin x}{x},$$

one can obtain the intensity distribution at the plane separated from the orifice plane by a distance z:

$$I(x,y,z) = I_0 \left(\frac{\sin kxa/z}{kxa/z}\right)^2 \left(\frac{\sin kyb/z}{kyb/z}\right)^2 \tag{2.70}$$

with

$$I_0 = \frac{a^2b^2k^2}{\pi^2 z^2}.$$

The same expression (see 2.43) was obtained above by using Kirchhoff's diffraction theory (see Fig. 2.2).

For slits with width $2a$ or $2b$ the last distribution will be transformed into

$$I(x) = I_0 \left(\frac{\sin kxa/z}{kxa/z}\right)^2 \tag{2.71}$$

and

$$I(y) = I_0 \left(\frac{\sin kyb/z}{kyb/z}\right)^2. \tag{2.72}$$

The distances between the first two zeros in these distributions are

$$\Delta x = \frac{\lambda z}{a} \tag{2.73}$$

and

$$\Delta y = \frac{\lambda z}{b}. \tag{2.74}$$

For a circular aperture with diameter D the amplitude transmittance is

$$t(r_1) = \mathrm{circ}\left(\frac{2r_1}{D}\right) \tag{2.75}$$

and the amplitude at the distance z from the aperture is defined as

$$\mathbf{U}(r) = C\mathcal{F}\{\mathbf{U}(r_1)\} \tag{2.76}$$

where

$$C = \frac{\exp(\mathrm{i}kz)}{\mathrm{i}\lambda z} \exp\left(\mathrm{i}\frac{kr^2}{2z}\right)$$

and r is a radius-vector in the observation plane, whereas r_1 is a radius-vector in the aperture plane. For collimated beam illumination, the intensity distribution in the Fraunhofer zone can be obtained as

$$I(r) = I_0 \left[\frac{2\mathcal{J}_1(kDr/z)}{kDr/z}\right]^2 . \tag{2.77}$$

This distribution, called the *Airy distribution*, was obtained also in Section 2.2 with the help of Kirchhoff's diffraction theory (see Eq. 2.46) and consists of a sequence of bright rings with rapidly decreasing amplitudes (see Fig. 2.2).

The radius-vector of the first dark ring is determined by the expression

$$\Delta r = 1.22\frac{\lambda z}{D}. \tag{2.78}$$

Let consider diffraction on the screen with two small apertures having diameters D and separated by a distance $2h$, e.g. along the ξ axis. Then the pupil function of the screen may be constructed as a convolution of $t(r_1)$, see Eq. (2.75), with delta-functions:

$$P(\xi,\eta) = t(r_1) \otimes [\delta(\xi + h, \eta) + \delta(\xi - h, \eta)]. \tag{2.79}$$

According to *the convolution theorem*, see [25], the Fourier transform of this pupil function is

$$\begin{aligned} \mathcal{F}\{P(\xi,\eta)\} &= \mathcal{F}\{t(r_1)\} \cdot [\exp(2\pi\mathrm{i}f_x h) + \exp(-2\pi\mathrm{i}f_x h)] \\ &= \pi\tfrac{D^2}{2} \cdot [2\tfrac{\mathcal{J}_1(kDr/z)}{kDr/z}] \cdot 2\cos(2\pi f_x h). \end{aligned} \tag{2.80}$$

Correspondingly, the intensity distribution in the Fraunhofer diffraction zone behind a screen with two small apertures is

$$I(x,y) = \mathrm{Const} \cdot [\frac{\mathcal{J}_1(kDr/z)}{kDr/z}]^2 \cdot [\cos(2\pi f_x h)]^2. \tag{2.81}$$

The diffraction zone behind a screen with two parallel slits having widths $2a$ and separated by a distance $2h$ along the ξ axis may be described in a similar manner and for the intensity distribution in the Fraunhofer zone one can obtain

$$I(x,y) = I_0 \cdot [\frac{\sin(kxa/z)}{kxa/z}]^2 \cdot [\cos(2\pi f_x h)]^2. \tag{2.82}$$

2.5 Diffraction at regular and chaotic structures

Let us consider now a screen consisting of N identical apertures. Under coherent illumination of the screen, the small apertures behave like coherent sources, and the amplitude of the wave diffracted at such a screen can be calculated as the sum of the amplitudes of the waves diffracted at the apertures. In the Fraunhofer approximation we have:

$$\begin{aligned} U(x,y) &= \text{Const} \sum_n \int \int_S \exp[-\mathrm{i}k\{(\xi_n + \xi)x + (\eta_n + \eta)y\}]\, \mathrm{d}\xi \mathrm{d}\eta \\ &= \text{Const} \sum_n \exp[-\mathrm{i}k\{x\xi_n + y\eta_n\} \\ &\quad \times \int \int_S \exp[-\mathrm{i}k\{x\xi + y\eta\}]\, \mathrm{d}\xi \mathrm{d}\eta \end{aligned} \tag{2.83}$$

where (ξ_1, η_1), (ξ_2, η_2), ..., (ξ_N, η_N) are the coordinates of similar points in different apertures related to the coordinate system spaced in the aperture plane and S is the area of one of the apertures.

If I_h is the intensity distribution in the diffraction zone for the case of one aperture (*the diffraction halo*), the resultant intensity can be calculated as

$$\begin{aligned} I(x,y) &= I_{\mathrm{h}}(x,y) |\sum_n \exp[-\mathrm{i}k(x\xi_n + y\eta_n)]|^2 \\ &= I_{\mathrm{h}}(x,y) \sum_n \sum_m |\exp[-\mathrm{i}k\{x(\xi_n - \xi_m) + y(\eta_n - \eta_m)\}]|^2 . \end{aligned} \tag{2.84}$$

The simplest case is diffraction at two apertures. For $N = 2$ we can obtain:

$$I = I_{\mathrm{h}} \cdot \cos^2(\frac{\Delta}{2z}), \tag{2.85}$$

where $\Delta = k[x(\xi_2 - \xi_1) + y(\eta_2 - \eta_1)]$. This relation is in accordance with Eq. (2.82) derived above in the previous section. Thus, the diffraction pattern $I_{\mathrm{h}}(x,y)$ is to be modulated by Young's interferometric fringes.

At the chaotic distribution of the apertures within the screen the terms with $m \neq n$ in Eq. (2.84) are rapidly oscillating in the interval between $+1$ and -1 and the sum of these terms tends to zero, when $N \to \infty$. The terms with $m = n$ are equal to 1 and according to this Eq. (2.84) transforms to

$$I(x,y) = N \cdot I_{\mathrm{h}}(x,y). \tag{2.86}$$

This result shows that the intensity distribution in the diffraction halo behind the screen with a chaotic distribution of N apertures is identical to the diffraction halo behind one aperture but with intensity N times as high.

Burch and Tokarski [1] confirmed that this result will be the same even in the case when the apertures are speckles recorded on photographic plate.

Let us continue the description of the diffraction at regular structures. It is easy to show (see [24]) that the diffraction at N equally spaced slits (an example of a grating) is described by the relation

$$I(x) = I_{\mathrm{h}}(x) \cdot \mathcal{H}(N, \frac{kxd}{z}) \tag{2.87}$$

where

$$\mathcal{H}(N, x) = (\frac{\sin Nx}{\sin x})^2 \tag{2.88}$$

and $2d$ is the grating period. For $N = 2$ Eq. (2.87) is reduced to Eq. (2.82)

The same relations with correspondingly modified $I_{\mathrm{h}}(x, y)$ will be valid for a grating consisting of N equally spaced apertures.

Figure 2.3 contains some examples of regular structures (gratings) the diffraction at which are often considered in speckle photography.

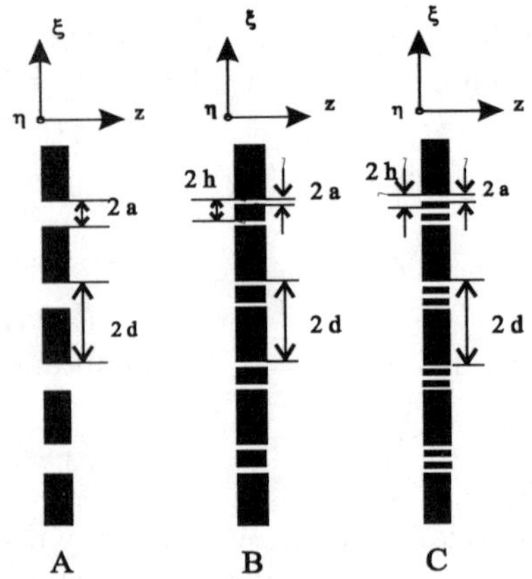

Fig. 2.3. Examples of 1D regular structures (gratings)

For the structure (A) on this figure we have

$$\begin{aligned} I(x) &= I_0 \cdot [\frac{\sin(Nkdx/z)}{\sin(kdx/z)}]^2 \cdot [\frac{\sin(kax/z)}{kax/z}]^2 \\ &= I_{\mathrm{h}} \cdot [\frac{\sin(Nkdx/z)}{\sin(kdx/z)}]^2 . \end{aligned} \tag{2.89}$$

For the cases (B) and (C) the term I_{h} in Eq. (2.89) should be modified:

$$I_{\mathrm{h}}(x) = I_0 \cdot [\frac{\sin(2khx/z)}{\sin(khx/z)}]^2 \cdot [\frac{\sin(kax/z)}{kax/z}]^2, \tag{2.90}$$

for the case (B) and

$$I_{\mathrm{h}}(x) = I_0 \cdot [\frac{\sin(3khx/z)}{\sin(khx/z)}]^2 \cdot [\frac{\sin(kax/z)}{kax/z}]^2, \tag{2.91}$$

for the case (C).

As we may see, with increasing N, the width of the interferometric fringes that modulate the diffraction halo is decreased and the distance between the maximum and minimum in the distribution (2.89) is

$$\Delta x = \frac{\lambda}{N2d}. \tag{2.92}$$

2.6 Diffraction in the image plane

Let us consider the optical geometry shown in Fig. 2.4. A monochromatic normally incident plane wave impinges upon a converging lens with focal length f. To construct the diffraction pattern in the image plane we will use a cascaded approach, considering the amplitude distributions subsequently at different planes and following [30].

Let us denote the amplitude distribution by $U(x, y)$ in the image plane, by $U_O(\xi_1, \eta_1)$ in an object plane, and by $U_L^-(\xi_2, \eta_2)$ and $U_L^+(\xi_2, \eta_2)$ in the lens plane just in front of the lens and immediately behind it, respectively. Let us first propagate the field amplitude from the plane (ξ_1, η_1) to the plane (ξ_2, η_2). According to the Fresnel approximation,

$$U_L^-(\xi_2, \eta_2) = \frac{\exp(\mathrm{i}kf)}{\mathrm{i}\lambda f} \exp\left[\mathrm{i}\frac{k}{2f}(\xi_2^2 + \eta_2^2)\right] \tag{2.93}$$

$$\times \int_{-\infty}^{+\infty}\int_{-\infty}^{+\infty} U_O(\xi_1, \eta_1) \exp\left[\mathrm{i}\frac{k}{2f}(\xi_1^2 + \eta_1^2)\right] \exp\left[-\mathrm{i}\frac{2\pi}{\lambda f}(\xi_1\xi_2 + \eta_1\eta_2)\right] \mathrm{d}\xi_1 \mathrm{d}\eta_1.$$

The propagation through the lens is described by a transmission function $T(\xi_2, \eta_2)$. This function is defined as follows. If an object with a transmission function $T(\xi_2, \eta_2)$ is illuminated by a plane wave with amplitude $U^-(\xi_2, \eta_2)$, then the amplitude just downstream of the object is given by

$$U^+(\xi_2, \eta_2) = U^-(\xi_2, \eta_2) \cdot T(\xi_2, \eta_2) \tag{2.94}$$

It was shown by Goodman [25] that the amplitude transmission function for the lens is calculated according to

$$T(\xi_2, \eta_2) = P(\xi_2, \eta_2) \cdot \exp\left[-\mathrm{i}\frac{k}{2f}(\xi_2^2 + \eta_2^2)\right], \tag{2.95}$$

where $P(\xi_2, \eta_2)$ is the lens pupil function. In the case of a circular lens we have $P(\xi_2, \eta_2) = 1$ for $\xi_2^2 + \eta_2^2 \leq D^2$ and $P(\xi_2, \eta_2) = 0$ elsewhere.

The propagation of the field from the lens to the back focal plane is calculated using the same Fresnel approximation:

$$U(x,y)=\frac{\exp(\mathrm{i}kf)}{\mathrm{i}\lambda f}\exp\left[\mathrm{i}\frac{k}{2f}(x^2+y^2)\right] \tag{2.96}$$

$$\times\int_{-\infty}^{+\infty}\int_{-\infty}^{+\infty}U_L^+(\xi_2,\eta_2)\exp\left[\mathrm{i}\frac{k}{2f}(\xi_1^2+\eta_1^2)\right]\exp\left[-\mathrm{i}\frac{2\pi}{\lambda f}(x\xi_2+y\eta_2)\right]\mathrm{d}\xi_2\mathrm{d}\eta_2.$$

Upon substitution of (2.95) into (2.94) the quadratic phase factor under the integral disappears and

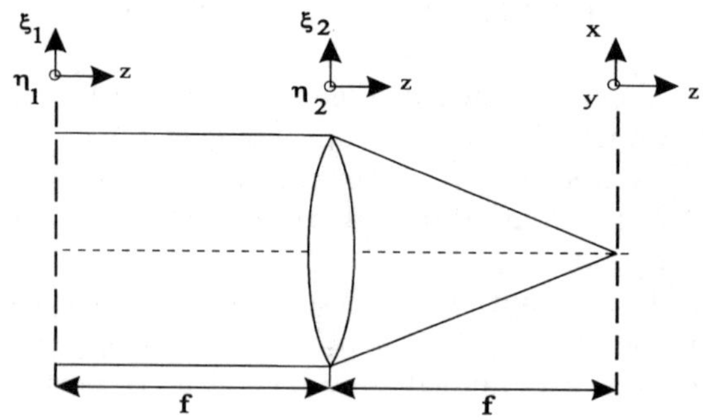

Fig. 2.4. Geometry for an optical Fourier transform

$$U(x,y)=\frac{\exp(\mathrm{i}kf)}{\mathrm{i}\lambda f}\exp\left[\mathrm{i}\frac{k}{2f}(x^2+y^2)\right]\cdot\mathcal{F}\{U_L^+(\xi_2,\eta_2)\} \tag{2.97}$$

where the Fourier transform of $U_L^+(\xi_2,\eta_2)$ is evaluated at frequencies $f_x=\frac{x}{\lambda f}$, $f_y=\frac{y}{\lambda f}$.

Finally, it is possible to express the amplitude distribution in the back focal plane of the lens as a Fourier transform of the amplitude distribution in the front focal plane of the lens [30]:

$$U(x,y)=\mathcal{F}\{U(\xi_1,\eta_1)\} = \frac{\exp(2\mathrm{i}kf)}{\mathrm{i}\lambda f}\int_{-\infty}^{\infty}\int_{-\infty}^{\infty}U(\xi_1,\eta_1) \times\exp\left[-\mathrm{i}\tfrac{2\pi}{\lambda f}(x\xi_1+y\eta_1)\right]\mathrm{d}\xi_1\mathrm{d}\eta_1 \tag{2.98}$$

Equation (2.98) shows that for an object placed in the front focal plane of the lens the amplitude distribution at the back focal plane is given by its Fourier transform. The transform must be evaluated at frequencies $f_x=\frac{x}{\lambda f}$ and $f_y=\frac{y}{\lambda f}$ to provide the proper scaling in the observation plane.

In the case of uniform illumination of the lens one can obtain

$$U(x,y) = \mathrm{Const}\cdot\mathcal{F}\{\mathcal{P}(\xi_2,\eta_2)\}. \tag{2.99}$$

The last function is called a *point spread function* and in the case of a circular lens is given by

$$\mathcal{H}(x,y) = \mathcal{F}\{\mathcal{P}(\xi_2,\eta_2)\} = \frac{2\mathcal{J}_1(\nu)}{\nu} \tag{2.100}$$

where $\mathcal{J}_1(\nu)$ is a first-order Bessel function of the first kind and

$$\nu = \frac{kr}{2f}. \tag{2.101}$$

Here r is the coordinate for this case of symmetry of revolution, $r^2 = x^2 + y^2$. Figure 2.2 shows the variation of intensity in the diffraction field, described by (2.100). It consists of a very luminous central spot, surrounded by alternating dark and bright rings.

Let us now consider the image formation in the geometry with the object and image distances equal to L_o and L_i, respectively. Denoting then by s the distance from a point in the object plane to a point in the lens plane, and by s' the distance from a point in the lens plane to a point in the image plane, the amplitude distribution in the image plane is given by [33]

$$\begin{aligned} U(x,y) &= \int_{-\infty}^{+\infty}\int_{-\infty}^{+\infty}\int_{-\infty}^{+\infty}\int_{-\infty}^{+\infty} U_\mathrm{O}(\xi_1,\eta_1)T(\xi_2,\eta_2) \\ &\times \frac{\exp(\mathrm{i}ks)}{s}\frac{\exp(\mathrm{i}ks')}{s'}\,\mathrm{d}\xi_1\mathrm{d}\eta_1\,\mathrm{d}\xi_2\mathrm{d}\eta_2. \end{aligned} \tag{2.102}$$

As in [33], the distances s and s' can be approximated in the following fashion:

$$s \simeq L_\mathrm{O} - \frac{\xi_1\xi_2 + \eta_1\eta_2}{L_\mathrm{O}} + \frac{\xi_2^2 + \eta_2^2}{2L_\mathrm{O}} \tag{2.103}$$

and

$$s' \simeq L_\mathrm{i} - \frac{x\xi_2 + y\eta_2}{L_\mathrm{i}} + \frac{\xi_2^2 + \eta_2^2}{2L_\mathrm{i}} \tag{2.104}$$

By substituting s and s' into Eq. (2.102) and using the lens formula

$$\frac{1}{L_\mathrm{O}} + \frac{1}{L_\mathrm{i}} = \frac{1}{f} \tag{2.105}$$

Khetan and Chiang [33] obtain

$$
\begin{aligned}
U(x,y) &= K \cdot \int_{-\infty}^{+\infty} \int_{-\infty}^{+\infty} \int_{-\infty}^{+\infty} \int_{-\infty}^{+\infty} U_{\mathrm{O}}(\xi_1,\eta_1) P(\xi_2,\eta_2) \\
&\quad \times \exp\left\{ \mathrm{i}k \left[\xi_2 \left(\frac{\xi_1}{L_{\mathrm{O}}} + \frac{x}{L_{\mathrm{i}}} \right) + \eta_2 \left(\frac{\eta_1}{L_{\mathrm{O}}} + \frac{y}{L_{\mathrm{i}}} \right) \right] \right\} \mathrm{d}\xi_1 \mathrm{d}\eta_1 \mathrm{d}\xi_2 \mathrm{d}\eta_2
\end{aligned}
\tag{2.106}
$$

where

$$
K = \frac{1}{L_{\mathrm{O}} L_{\mathrm{i}}} \exp[-\mathrm{i}k(L_{\mathrm{O}} + L_{\mathrm{i}})] \tag{2.107}
$$

and $P(\xi_2, \eta_2)$ is the lens pupil function.

3. Statistical Properties of Speckle Patterns

The purpose of this chapter is to introduce the various basic statistical properties of speckle patterns. We will basically follow Goodman [22] with some additions concerning *dynamic speckle* behaviour. Two main functions, namely, the probability density functions (PDF) and correlation functions (CFs) will be introduced and used for the statistical description of a speckle pattern.

In general the statistical properties of a speckle pattern depend both on the coherence of the light forming speckles and on the detailed properties of the speckle-forming medium. As the first approximation, we describe here the statistics for perfectly coherent light and for scatters that introduce path differences greater than one wavelength. For this case the speckle statistics do not depend on the above mentioned factors.

In spite of the fact that speckle phenomena have been investigated since Newton's time (for a review see [5]), the term 'speckle' was introduced to describe the granular structure of objects under laser illumination relatively recently, when the first He-Ne lasers became available, see for instance [22,34].

3.1 The probability density function

Let us start from the first order statistics which describes the statistical properties of speckles at a single point and assume that the light is perfectly monochromatic, polarized and coherent. We may decompose the complex amplitude introduced by Eq. (2.13) into the real and imaginary parts of the speckle field

$$\mathbf{U} = \mathrm{Re}(\mathbf{U}) + \mathrm{i}\ \mathrm{Im}(\mathbf{U}). \tag{3.1}$$

Then

$$I = [\mathrm{Re}(\mathbf{U})]^2 + [\mathrm{Im}(\mathbf{U})]^2, \quad \text{and} \quad \Theta = \tan^{-1}\frac{\mathrm{Im}(\mathbf{U})}{\mathrm{Re}(\mathbf{U})} \tag{3.2}$$

are the speckle intensity and the associated phase angle.

For all optical systems of speckle-pattern formation, the light intensity at the given point results from the 3D coherent addition of many de-phased independent contributions. Thus, the complex amplitude of the resultant field at the observation point $\mathbf{U}(x, y, z)$ is represented as a sum of many elementary contributions

$$\mathbf{U}(x, y, z) = \sum_{k=1}^{N} \frac{1}{\sqrt{N}} \mathbf{U}_k(x, y, z) = \frac{1}{\sqrt{N}} \sum_{k=1}^{N} |U_k| \cdot \exp(\mathrm{i}\theta_k). \tag{3.3}$$

We may write the same relation for the real and imaginary parts of the resultant field

$$\mathrm{Re}(\mathbf{U})(x, y, z) = \sum_{k=1}^{N} \frac{1}{\sqrt{N}} \mathrm{Re}(\mathbf{U}_k)(x, y, z) = \frac{1}{\sqrt{N}} \sum_{k=1}^{N} |U_k| \cdot \cos\theta, \tag{3.4}$$

$$\mathrm{Im}(\mathbf{U})(x, y, z) = \sum_{k=1}^{N} \frac{1}{\sqrt{N}} \mathrm{Im}(\mathbf{U}_k)(x, y, z) = \frac{1}{\sqrt{N}} \sum_{k=1}^{N} |U_k| \cdot \sin\theta. \tag{3.5}$$

The average values of $\mathrm{Re}(\mathbf{U})$ and $\mathrm{Im}(\mathbf{U})$ over an ensemble[1] of macroscopically similar but microscopically different speckle-forming media are

$$< \mathrm{Re}(\mathbf{U}) >= \frac{1}{\sqrt{N}} \sum_{k=1}^{N} < |U_k| \cdot \cos\theta >= \frac{1}{\sqrt{N}} \sum_{k=1}^{N} < |U_k| > \cdot < \cos\theta >= 0, \tag{3.6}$$

and

$$< \mathrm{Im}(\mathbf{U}) >= \frac{1}{\sqrt{N}} \sum_{k=1}^{N} < |U_k| \cdot \sin\theta >= \frac{1}{\sqrt{N}} \sum_{k=1}^{N} < |U_k| > \cdot < \sin\theta >= 0. \tag{3.7}$$

We have use here the assumptions that the amplitudes and phases of the elementary phasors are statistically independent, and the phases are uniformly distributed on the primary interval $(-\pi, \pi)$.

Under these assumptions and using the *central limit theorem* at $N \to \infty$, it is possible to show[2] that $\mathrm{Re}(\mathbf{U})$ and $\mathrm{Im}(\mathbf{U})$ are asymptotically Gaussian. Taking into account this fact Goodman [22] obtains an expression for the joint probability density function (PDF) of the real and imaginary parts of the resultant field:

$$p\{\mathrm{Re}(\mathbf{U}), \mathrm{Im}(\mathbf{U})\} = \frac{1}{2\pi\sigma^2} \exp\left\{-\frac{[\mathrm{Re}(\mathbf{U})]^2 + [\mathrm{Im}(\mathbf{U})]^2}{2\sigma^2}\right\} \tag{3.8}$$

---

[1] An ensemble average could be replaced by a spatial one

[2] From the mathematical point of view this problem is identical to the classical statistical problem of a random walk in a plane, see [22].

where

$$\sigma^2 = \lim_{N\to\infty} \frac{1}{N} \sum_{k=1}^{N} \frac{<|U_k|^2>}{2} = \frac{<I>}{2} \tag{3.9}$$

and $< \dots >$ denote values average over an ensemble of macroscopically similar but microscopically different speckle patterns.

This function is a circular Gaussian function and contours of constant probability density are circles in the complex plane.

For most speckle photography experiments, it is the intensity of the wave that is directly measured. The joint PDF of the intensity of the speckle field and its phase is determined by

$$p_{I,\Theta}(I,\Theta) = \begin{cases} \frac{1}{2\pi<I>} \exp\left(-\frac{I}{<I>}\right), & \text{when } I \geq 0; -\pi \leq \Theta < \pi \\ 0 & \text{otherwise.} \end{cases} \tag{3.10}$$

Here $p_{I,\Theta}(I,\Theta)\mathrm{d}I\mathrm{d}\Theta$ is the probability that light in a speckle has intensity (brightness) between the values I and $I + \mathrm{d}I$ and phase between the values Θ and $\Theta + \mathrm{d}\Theta$.

According to these assumptions,

$$p(I,\Theta) = p(I) \cdot p(\Theta), \tag{3.11}$$

therefore

$$p(I) = \int_{-\pi}^{\pi} p(I,\Theta)\,\mathrm{d}\Theta = \frac{1}{<I>} \exp(-\frac{I}{<I>}), \tag{3.12}$$

and

$$p(\Theta) = \int_{0}^{\infty} p(I,\Theta)\,\mathrm{d}I = \frac{1}{2\pi}. \tag{3.13}$$

These relations show that the intensity obeys negative exponential statistics, while the phase obeys uniform statistics.

The nth moment $< I^n >$ of the intensity is equal to

$$< I^n >= n!(2\sigma^2)^n = n! < I >^n, \tag{3.14}$$

and for the special case $n = 1$ we have $< I >= 2\sigma^2$.

The standard deviation of the intensity of a speckle pattern is

$$\sigma_I^2 =< I^2 > - < I >^2 =< I >^2 . \tag{3.15}$$

The ratio of the standard deviation to the mean intensity may be defined as the *contrast* of the speckle pattern:

$$C = \frac{\sigma_I}{< I >} \tag{3.16}$$

and from this definition we see that the contrast is equal to unity in this particular case.

It is convenient to introduce the probability $P(I)$ that the intensity exceeds a threshold I

$$P(I) = \int_I^\infty \frac{1}{< I >} \exp(-\frac{\xi}{< I >}) d\xi = \exp(-\frac{I}{< I >}). \tag{3.17}$$

It is interesting to note that the most probable intensity at any point in such a speckle pattern is zero, i.e. there are more dark speckles in the field than speckles of any other brightness.

The negative exponential PDF for the intensity in a speckle field was first derived by Goodman, and later the results of experimental verification based on 23 000 measurements of intensity have shown good agreement between theory and experiment; see [22].

It was shown that coherent addition of speckle patterns, on an amplitude basis, does not change the statistics of the intensity, apart from a scaling constant. Contrary to this, in the case of non-coherent addition, the resultant intensity of a speckle pattern is expressed as a sum not of amplitudes but of intensities:

$$I = \sum_{k=1}^{N} I_k. \tag{3.18}$$

The statistics of such a speckle pattern depend on the correlation coefficients describing the correlations between the N amplitude and intensity components:

$$\mu_{kl} = \frac{< \mathbf{U}_k \mathbf{U}_l^* >}{[< |\mathbf{U}_k|^2 >< |\mathbf{U}_l|^2 >]^{\frac{1}{2}}} \tag{3.19}$$

$$c_{kl} = \frac{< I_k I_l > - < I_k >< I_l >}{[< (I_k - < I_k >)^2 >< (I_l - < I_l >)^2 >]^{\frac{1}{2}}}. \tag{3.20}$$

These correlation coefficients are related by

$$c_{kl} = |\mu_{kl}|^2 \tag{3.21}$$

and

$$< I_k I_l > = < I_k >< I_l > [1 + |\mu_{kl}|^2]. \tag{3.22}$$

The results of calculations of the PDFs for the fields composed from the fields with different correlations show that the contrast of the resultant speckle field reaches its minimum value of $1/\sqrt{2}$ when the summing fields are fully decorrelated.

For the particular case of a field composed from a sum of two fields with equal average intensities, one of which is a speckle field and the second is a uniform coherent field, the PDF of the intensity distribution is [23]

$$p(I) = \frac{2}{<I>} \exp[-\frac{<I>+2I}{<I>}] \cdot \mathcal{J}_0(2\sqrt{\frac{2I}{<I>}}) \qquad (3.23)$$

where $\mathcal{J}_0$ is the Bessel function of the first kind of zero order. This distribution is shown in Fig. 3.1.[3]

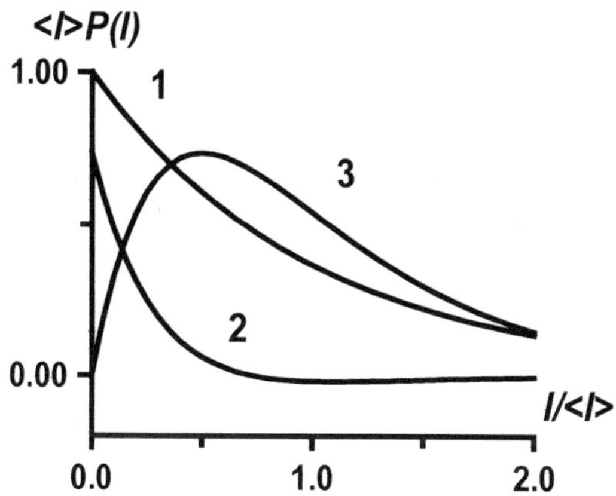

Fig. 3.1. Normalized probability density functions of the intensity distribution of speckle fields. (1) Single speckle field, (2) coherent sum of speckle field and uniform field, (3) incoherent sum of two speckle fields

As we can see, this distribution does not differ greatly from the distribution curve for speckle field alone. In particular, the most probable intensity is still zero.

The result of the coherent addition of two speckle patterns is a third speckle field, differing in detail from its two constituent patterns, but whose statistical distributions remain the same. For incoherent addition of the two speckle patterns the intensity distribution follows the equation [23]

$$p(I) = \frac{4I}{<I>^2} \cdot \exp[-\frac{2I}{<I>}]. \qquad (3.24)$$

This distribution is also plotted in Fig. 3.1. It is seen that now there is a very low probability of a dark speckle occurring. This is because there is a high probability that the bright areas of one pattern will be superimposed on to the dark areas of the other. Such incoherent addition of two speckle patterns

[3] In [23] it is noted, that a matt white paint surface or oral glass, into which the light penetrates and is multiply scattered, depolarizes the light and thus does not generate a fully developed speckle pattern. Ennos has shown that the brightness distribution of this type of speckle field is markedly different from that of the 'fully-developed' speckle pattern.

could be obtained, e.g. under reflection of monochromatic light from the surfaces that completely depolarize the light [23], since any two orthogonally polarized components of the scattered light are incoherent with one another.

Comparing the distributions of Eq. (3.23) and Eq. (3.24) one may conclude that there is a striking statistical difference between the intensity distribution of coherently and incoherently summed speckle patterns. The practical ways of quantitative characterization of this difference will be discussed in chapter 5.

The reader ca find a more detailed description of the PDF's of the summed speckle pattern in [22].

3.2 The spatial structure of the speckle field

To describe the spatial structure of a speckle pattern[4] one needs to use the second-order statistical properties of the speckle.

Consider the autocorrelation function of the speckle intensity distribution in some plane orthogonal to the direction of light propagation. According to its definition the autocorrelation function may be evaluated as

$$R_I(x_1, y_1; x_2, y_2) = < I(x_1, y_1) I(x_2, y_2) > . \tag{3.25}$$

Here the average is over an ensemble of rough surfaces. The width of this autocorrelation function provides a reasonable measure of the average speckle size.

Introducing vectors $\mathbf{r}_1 = \mathbf{r}_1(x_1, y_1)$ and $\mathbf{r}_2 = \mathbf{r}_2(x_2, y_2)$, the autocorrelation function is

$$R_I(\mathbf{r}_1, \mathbf{r}_2) = < I(\mathbf{r}_1) I(\mathbf{r}_2) > . \tag{3.26}$$

For homogeneous random fields the correlation function depends only on the vector $\mathbf{r} = \mathbf{r}_2 - \mathbf{r}_1$. This means that the correlation function of a homogeneous field does not change when the pair of points $\mathbf{r}_1$ and $\mathbf{r}_2$ are both displaced by the same amount in the same direction, i.e.

$$R_I(\mathbf{r}_1, \mathbf{r}_2) = R_I(\mathbf{r}_1 + \mathbf{r}_0, \mathbf{r}_2 + \mathbf{r}_0).$$

The mean values for such fields are constants

$$< I >= \text{Const}; \quad < \mathbf{U} >= \text{Const}.$$

A homogeneous random field is called isotropic if its intensity correlation function $R_I(\mathbf{r})$ depends only on $r = |\mathbf{r}|$.

The intensity autocorrelation function can be expressed through the autocorrelation function of the amplitudes, which we represent by

[4] Goodman has using the term *coarseness* of the speckle spatial structure, [22].

$$R_{\mathbf{U}}(\mathbf{r}_1, \mathbf{r}_2,) = < \mathbf{U}(\mathbf{r}_1)\mathbf{U}^*(\mathbf{r}_2) > . \tag{3.27}$$

For circular complex Gaussian fields, the required relation between R_I and $R_{\mathbf{U}}$ is [22]

$$R_I(\mathbf{r}_1, \mathbf{r}_2) = < I(\mathbf{r}_1) >< I(\mathbf{r}_2) > + |R_{\mathbf{U}}(\mathbf{r}_1, \mathbf{r}_2)|^2. \tag{3.28}$$

The autocorrelation function of the amplitudes, $R_{\mathbf{U}}$, is also referred to as the mutual intensity of the field. In many cases, it is convenient to deal with a normalized version of the mutual intensity known as the complex coherence factor and defined by

$$\mu_{\mathbf{U}}(\mathbf{r}_1, \mathbf{r}_2) = \frac{R_{\mathbf{U}}(\mathbf{r}_1, \mathbf{r}_2)}{[R_{\mathbf{U}}(x_1, y_1; x_1, y_1)R_{\mathbf{U}}(x_2, y_2; x_2, y_2)]^{\frac{1}{2}}}. \tag{3.29}$$

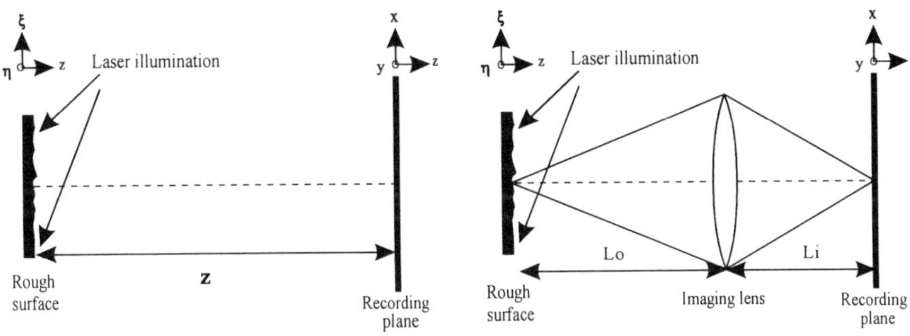

Fig. 3.2. Speckle formation using free-space propagation geometry (left) and image formation geometry (right)

Let us consider the free-space propagation geometry shown in Fig. 3.2. The speckle patterns are formed as a result of random 3D interference of the monochromatic coherent light diffusely scattered from a rough surface. Following Goodman [22], the calculation of the mutual intensity in the observation plane is begun using the Huygens–Fresnel principle, expressed in the Fresnel approximation as (see Eq. 2.29)

$$\begin{aligned} U(x,y) &= \tfrac{1}{\lambda z} \exp\left[-\mathrm{i}\tfrac{\pi}{\lambda z}(x^2+y^2)\right] \int\int_{-\infty}^{\infty} U_s(\xi,\eta) \\ &\quad \times \exp\left[-\mathrm{i}\tfrac{\pi}{\lambda z}(\xi^2+\eta^2)\right] \exp\left[\mathrm{i}\tfrac{2\pi}{\lambda z}(x\xi+y\eta)\right] \mathrm{d}\xi \mathrm{d}\eta. \end{aligned} \tag{3.30}$$

Here $U_s(\xi, \eta)$ is the complex amplitude at the scattering surface and $U(x, y)$ is the amplitude of the scattered field in the observation plane.

In general form the mutual intensity J_U for this case can be expressed as

$$\begin{aligned} J_U(\mathbf{r}_1,\mathbf{r}_2) &= \tfrac{1}{\lambda^2 z^2}\exp\left[-\mathrm{i}\tfrac{\pi}{\lambda z}(x_1^2-x_2^2+y_1^2-y_2^2)\right] \\ &\times \int\int\int\int_{-\infty}^{\infty} J_{U_s}(\xi_1,\eta_1;\xi_2,\eta_2)\exp\left[-\mathrm{i}\tfrac{\pi}{\lambda z}(\xi_1^2-\xi_2^2+\eta_1^2-\eta_2^2)\right] \\ &\times \exp\left[\mathrm{i}\tfrac{2\pi}{\lambda z}(x_1\xi_1+y_1\eta_1-x_2\xi_2-y_2\eta_2)\right]\mathrm{d}\xi_1\mathrm{d}\eta_1\mathrm{d}\xi_2\mathrm{d}\eta_2. \end{aligned} \tag{3.31}$$

Assuming a simple relation between $J_{U_s}(\xi_1,\eta_1;\xi_2,\eta_2)$ and the amplitude of the field incident on the scattered spot, $U_i(\xi,\eta)$, of the form:

$$J_{U_s}(\xi_1,\eta_1;\xi_2,\eta_2) = kU_i(\xi_1,\eta_1)U_i^*(\xi_2,\eta_2)\delta(\xi_1-\xi_2,\eta_1-\eta_2) \tag{3.32}$$

where $\delta(\xi,\eta)$ is the two-dimensional delta function and k is the proportionality constant, it is possible to simplify Eq. (3.31):

$$\begin{aligned} J_U(\mathbf{r}_1,\mathbf{r}_2) &= \tfrac{k}{\lambda^2 z^2}\int\int_{-\infty}^{\infty}|U_s(\xi_1,\eta_1)|^2 \\ &\times \exp\left[\mathrm{i}\tfrac{2\pi}{\lambda z}\left\{\xi_1(x_1-x_2)+\eta_1(y_1-y_2)\right\}\right]\mathrm{d}\xi_1\mathrm{d}\eta_1. \end{aligned} \tag{3.33}$$

The autocorrelation function of the speckle intensity assumes the form

$$R_I(\Delta x,\Delta y) =< I >^2 \left[1+|\mu_U(\Delta x,\Delta y)|^2\right] \tag{3.34}$$

$$=< I >^2 \left[1+\left|\frac{\int\int_{-\infty}^{\infty}|U_s(\xi,\eta)|^2\exp\left[\mathrm{i}\frac{2\pi}{\lambda z}\left\{\xi\Delta x+\eta\Delta y\right\}\right]\mathrm{d}\xi\mathrm{d}\eta}{\int\int_{-\infty}^{\infty}|U_s(\xi_1,\eta_1)|^2\mathrm{d}\xi\mathrm{d}\eta}\right|^2\right].$$

or, in the special case of a uniform and square scattering spot with dimensions $2a \times 2b$, we have

$$|U(\xi,\eta)|^2 = \mathrm{rect}\frac{\xi}{2a}\mathrm{rect}\frac{\eta}{2b}, \tag{3.35}$$

where $\mathrm{rect}x = 1$ for $|x| \leq \frac{1}{2}$ or zero otherwise. The autocorrelation function of the intensity is then given by

$$R_I(\Delta x,\Delta y) =< I >^2 \left[1+\mathrm{sinc}^2\frac{2a\Delta x}{\lambda z}\mathrm{sinc}^2\frac{2b\Delta y}{\lambda z}\right]. \tag{3.36}$$

The width of the autocorrelation function provides a reasonable measure of the average width of a speckle, 'speckle size'. This speckle size, σ_{sp}, for the latter case can be taken to be the values of $\Delta x, \Delta y$, where $\mathrm{sinc}^2(2a\Delta x/\lambda z)$ and $\mathrm{sinc}^2(2b\Delta y/\lambda z)$ first falls to zero. Denoting these distances by δx and δy, we have

$$\sigma_{\mathrm{sp}_x} = \delta x = \frac{\lambda z}{2a}, \quad \sigma_{\mathrm{sp}_y} = \delta y = \frac{\lambda z}{2b}. \tag{3.37}$$

Having a correlation function, it is possible to evaluate the *power spectral density* of the speckle intensity distribution $I(x, y$. According to the Wiener–Khintchine theorem (see [22]), the *power spectral density* $\mathcal{W}(f_x, f_y)$ is given by the Fourier transformation of the intensity autocorrelation function

$$\mathcal{W}(f_x, f_y) = |\mathcal{F}\{I(x,y)\}|^2 = \mathcal{F}\{R_I(x,y)\}. \tag{3.38}$$

The power spectral density is also called *the Wiener spectrum.*

For the particular case of a uniform and square scattering spot, the power spectral density is of the form [22]

$$\mathcal{W}(f_x, f_y) =< I >^2 \left[\delta(f_x, f_y) + \left(\frac{\lambda z}{L}\right)^2 \cdot \Lambda\left(\frac{\lambda z f_x}{L}\right) \cdot \Lambda\left(\frac{\lambda z f_y}{L}\right)\right] \tag{3.39}$$

where $L = 2a = 2b$ and $\Lambda(x) = 1 - |x|$ for $|x| \leq 1$, zero otherwise. As we can see, the speckle pattern contains no frequency components higher than $\frac{L}{\lambda z}$, cycles/m both in the f_x and in the f_y components.

Let us now go to the analysis of autocorrelation functions for the case involving imaging systems. This case is also illustrated in Fig. 3.2. We suppose that the object of concern is uniformly reflective and the illuminated region is broad as compared with the resolution cell associated with the lens used. For the usual case of a circular lens pupil of diameter D, the autocorrelation function of the image-plane speckle field is

$$R_I(r) =< I >^2 \left[1 + |2\frac{\mathcal{J}_1(\frac{\pi D r}{\lambda L_i})}{\frac{\pi D r}{\lambda L_i}}|^2\right] \tag{3.40}$$

where $r = [(\Delta x)^2 + (\Delta y)^2]^{\frac{1}{2}}$.

The speckle size for this case can be taken to be the value of Δr, where the Bessel function of the first kind and first order, $\mathcal{J}_1$, first falls to zero. Denoting this distance by δr, we have

$$\sigma_{\text{sp}} = \delta r = 1.22\frac{\lambda L_i}{D}. \tag{3.41}$$

These correlation distances are in good accordance with the estimations obtained by using diffraction theory for single apertures.

The power spectral density of the speckle pattern in the image plane for the usual case of a circular lens of diameter D is of the form [22]

$$\begin{aligned} \mathcal{W}(\rho) &= < I >^2 \Big\{\delta(\rho) + \left(\tfrac{\lambda L_i}{D}\right)^2 \\ &\times \tfrac{4}{\pi} \cdot \left[\cos^{-1}\left(\tfrac{\lambda L_i \rho}{D}\right) - \tfrac{\lambda L_i \rho}{D}\sqrt{1 - \left(\tfrac{\lambda L_i \rho}{D}\right)^2}\right]\Big\} \end{aligned} \tag{3.42}$$

for $\rho \leq D/\lambda L_i$, and zero otherwise, where $\rho = \sqrt{f_x^2 + f_y^2}$.

3.3 Dynamic speckle statistics

Random interference of laser light scattered from a moving diffuse object produces a speckle pattern which varies in time and space. Such speckles are called *dynamic laser speckles*, see [35]. The most general tool for investigating the statistical properties of dynamic speckles is to use the space-time correlation function of the speckle intensity, as proposed by Anisimov *et al.*, see [36].

Let the diffuse object be a deep random phase screen moving at the object plane. Depending on the position of the observation plane, two different fundamental motions of speckles are produced. The first type is when the speckles move as a whole and their shape remains unchanged for a considerable displacement. This type of speckle motion is called a 'translation'[5].

The second type is when the individual speckles deform, disappear, and reappear without appreciable displacement of their positions. This type of speckle motions is called 'boiling' of speckles.[6]

In many practical cases the dynamic behaviour of speckles is composed of both boiling and translation, which behaviour depends on the conditions of both the optical configuration and the illumination light. Yoshimura has shown that the phenomenon of speckles boiling is affected by, in addition to the above two conditions, the Doppler-shift frequency difference of the light scattered within the point spread of the optical system used [37].

Analysing the space-time correlation functions of dynamic speckles, Anisimov *et al.* [36] have shown that the translation and boiling motions of dynamic speckles occur under the near- and far-field conditions, respectively.[7]

3.3.1 Statistical properties of dynamic speckles in the diffraction field

Consider the statistics of time-varying speckles produced in the far-field region of an illuminated area of the moving diffuse object, see Fig. 3.3.

The Cartesian coordinates of the positions at the object and observation planes are denoted by $\mathbf{r}_0 = \mathbf{r}_0(\xi, \eta)$ and $\mathbf{r} = \mathbf{r}(x, y)$, respectively.

When the diffuse object moving with velocity $\mathbf{v} = \mathbf{v}(v_\xi, v_\eta)$ at the object plane is illuminated by monochromatic light with complex amplitude $\mathbf{U}_-(\mathbf{r}_0)$, the amplitude distribution of light emerging from the diffuse object may be written as (see [35])

$$\mathbf{U}_+(\mathbf{r}_0, t) = \mathbf{U}_-(\mathbf{r}_0)\exp[\mathrm{i}\Theta(\mathbf{r}_0 - \mathbf{v}t)]. \tag{3.43}$$

[5] This type of speckle motion is similar to turbulent flow with a 'frozen' turbulence, when molecular diffusion is small compared to convection.

[6] For the case of turbulence such motion corresponds to very small-scale turbulence with significant dissipation.

[7] In general case the type of the speckle motion depends on the condition of illumination, see [37].

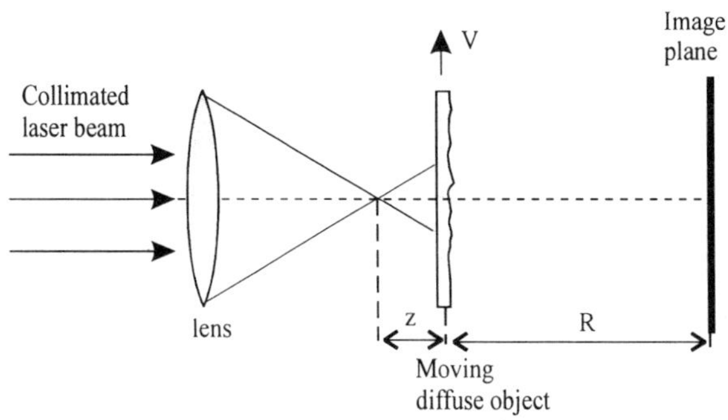

Fig. 3.3. Speckle formation by moving diffuse object

Here $\Theta(\xi, \eta)$ is the phase variation due to the random phase screen

The amplitude distribution of the speckle field at the observation plane in the diffraction field , a distance R away from the object plane, is given by

$$\mathbf{U}(\mathbf{r}, t) = \int\int_{(S)} \mathbf{U}_{+}(\mathbf{r}_0, t)\mathbf{K}(\mathbf{r}_0, \mathbf{r})\, \mathrm{d}\mathbf{r}_0, \tag{3.44}$$

where $\mathbf{K}(\mathbf{r}_0, \mathbf{r})$ is the propagation function of the light amplitude from the object to the observation plane.

For Fresnel diffraction this propagation function, according to Eq. (2.22) is equal to

$$\mathbf{K}(\mathbf{r}_0, \mathbf{r}) = \exp\left(\frac{\mathrm{i}\pi|\mathbf{r}_0 - \mathbf{r}|^2}{\lambda R}\right). \tag{3.45}$$

The speckle intensity detected at the point $\mathbf{r}$ is denoted by $I(\mathbf{r}, t) = |\mathbf{U}(\mathbf{r}, t)|^2$.

The space-time correlation function of the complex amplitude $\mathbf{U}(\mathbf{r}, t)$ is written as

$$R_{\mathbf{U}}(\mathbf{r}_1, \mathbf{r}_2; t_1, t_2) =< \mathbf{U}(\mathbf{r}_1, t_1)\mathbf{U}^*(\mathbf{r}_2, t_2) > . \tag{3.46}$$

For an object moving with constant velocity this autocorrelation function will be stationary and will depend on the time interval $\tau = (t_2 - t_1)$.

Introducing the correlation area of the wave phase distribution on the diffuse object, ΔS_{r_0}, Asacura and Tokai [35] have obtained a general equation for the space-time correlation function

$$\begin{aligned} R_{\mathbf{U}}(\mathbf{r}_1, \mathbf{r}_2; \tau) &= \Delta S_{r_0} \int\int_{(S)} \mathbf{U}_{+}(\mathbf{r}_0 - \tfrac{\mathbf{v}}{2}\tau)\mathbf{U}_{+}^*(\mathbf{r}_0 + \tfrac{\mathbf{v}}{2}\tau) \\ &\quad \times \mathbf{K}\left(\mathbf{r}_0 - \tfrac{\mathbf{v}}{2}\tau, \mathbf{r}_1\right) \mathbf{K}^*\left(\mathbf{r}_0 + \tfrac{\mathbf{v}}{2}\tau, \mathbf{r}_2\right) \mathrm{d}\mathbf{r}_0. \end{aligned} \tag{3.47}$$

Use of Eq. (3.45) and Eq. (3.47) yields

$$\begin{aligned} R_{\mathbf{U}}(\mathbf{r}_1, \mathbf{r}_2; \tau) &= \Delta S_{r_0} \exp[\mathrm{i}\tfrac{\pi}{\lambda R}(\mathbf{r}_1 + \mathbf{r}_2)\mathbf{v}\tau] \\ &\quad \times \exp[\mathrm{i}\tfrac{\pi}{\lambda R}(\mathbf{r}_1^2 - \mathbf{r}_2^2)] \\ &\quad \times \textstyle\int\int_{(S)} \mathbf{U}_+(\mathbf{r}_0 - \frac{\mathbf{v}}{2}\tau)\mathbf{U}_+^*(\mathbf{r}_0 + \frac{\mathbf{v}}{2}\tau) \\ &\quad \times \exp\{\mathrm{i}\tfrac{2\pi}{\lambda R}[(\mathbf{r}_2 - \mathbf{r}_1) - \mathbf{v}\tau]\mathbf{r}_0\}\, \mathrm{d}\mathbf{r}_0. \end{aligned} \tag{3.48}$$

For a homogeneous and stationary speckle pattern the complex coherence factor, defined by Eq. (3.19), is given by

$$\begin{aligned} \mu_{\mathbf{U}}(\mathbf{x}, \tau) &= \tfrac{|R_{\mathbf{U}}(\mathbf{x},\tau)|}{|R_{\mathbf{U}}(0,0)|} \\ &= C \cdot |\textstyle\int\int_{(S)} \mathbf{U}_+(\mathbf{r}_0 - \frac{\mathbf{v}}{2}\tau)\mathbf{U}_+^*(\mathbf{r}_0 + \frac{\mathbf{v}}{2}\tau)| \\ &\quad \times \exp\{\mathrm{i}\tfrac{2\pi}{\lambda R}[(\mathbf{x} - \mathbf{v}\tau)]\mathbf{r}_0\}\, \mathrm{d}\mathbf{r}_0, \end{aligned} \tag{3.49}$$

where

$$C = \{|\int\int_{(S)} |\mathbf{U}_+(\mathbf{r}_0)|^2 \mathrm{d}\mathbf{r}_0|\}^{-1}$$

and

$$\mathbf{x} = \mathbf{r}_2 - \mathbf{r}_1.$$

By putting $\tau = 0$ in Eq. (3.49), the complex coherence factor is simplified to

$$\mu_{\mathbf{U}}(\mathbf{x}, 0) = C \cdot |\int\int_{(S)} |\mathbf{U}_+(\mathbf{r}_0)|^2 \exp\{\mathrm{i}\frac{2\pi}{\lambda R}(\mathbf{x}\mathbf{r}_0)\}\, \mathrm{d}\mathbf{r}_0. \tag{3.50}$$

This expression means that the space correlation function of the speckle intensity fluctuation, in the Fresnel diffraction zone, is given by a Fourier transform of the intensity distribution of the illuminating light. This relation is quite identical in form to the van Cittert–Zernike theorem which indicates that the spatial coherence in the far-field region of an incoherent source is given by taking the Fourier transform of the source intensity distribution [35].

The normalized space correlation function of the speckle intensity fluctuation in *a static speckle pattern* ($\tau = 0$) is given by

$$\frac{R_{\Delta I}(\mathbf{x}, 0)}{R_{\Delta I}(0, 0)} = |\mu_{\mathbf{U}}(\mathbf{x}, 0)|^2 = \exp\left(-\frac{|\mathbf{x}|^2}{|\Delta x|^2}\right), \tag{3.51}$$

where

$$\Delta x = \frac{\lambda R}{\pi w},$$

and w is the width of the beam at the object plane. The value of Δx here is related to the speckle size in the diffraction field.

For single point observation, $\mathbf{x} = 0$, the normalized *time* autocorrelation function is written as

$$\frac{R_{\Delta \mathrm{I}}(0,\tau)}{R_{\Delta \mathrm{I}}(0,0)} = |\mu_{\mathrm{U}}(0,\tau)|^2 = \exp\left(-\frac{|\tau|^2}{|\tau_{\mathrm{C}}|^2}\right), \tag{3.52}$$

where τ_{C} is the time correlation length of *the dynamic* (time varying) speckle intensity, defined as

$$\tau_{\mathrm{C}} = \frac{1}{\mathbf{v}}\left(\frac{1}{w^2} + \frac{\sigma^2}{\Delta x^2}\right)^{-\frac{1}{2}} \tag{3.53}$$

where

$$\sigma = \frac{R}{\rho} + 1 \tag{3.54}$$

with ρ being the wavefront-curvature radius and ω being the width of the beam in the object plane.

Thus the time correlation length τ_{C} of the time-varying speckle intensity is inversely proportional to the absolute value of the velocity of a moving object and the proportionality constant is intimately related to the average size of speckles and the illumination conditions of the Gaussian beam through the parameters of w and ρ [35].

3.3.2 Time and spatial averaging of speckle pattern

The measured speckle intensity is spatially integrated when the spatial resolution of the detector is finite. The same effect is observed if the speckle pattern is blurred, as for single (prolonged) exposure speckle photography, see chapter 5.

Thus, if a speckle pattern with intensity distribution $I(x,y)$ is detected through an extended aperture, the measured intensity is (see [22])

$$I_0 = \frac{1}{S}\int_{-\infty}^{\infty}\int_{-\infty}^{\infty} \mathcal{S}(x,y) I(x,y)\,\mathrm{d}x\mathrm{d}y \tag{3.55}$$

with $\mathcal{S}(x,y)$ being a weighting function and

$$S = \int_{-\infty}^{\infty}\int_{-\infty}^{\infty} \mathcal{S}(x,y)\,\mathrm{d}x\mathrm{d}y.$$

For a detector having a uniform response over a finite aperture, the weighting function has the simple form

$$\mathcal{S}(x,y) = \begin{cases} 1, & \text{in the aperture} \\ 0, & \text{outside the aperture,} \end{cases} \tag{3.56}$$

and S is the area of that aperture.

Similarly, for the intensity measured when the speckle pattern is in motion, the weighting function depends on the integration time. For the speckle movement with a constant velocity v along the y-axis, the weighting function takes the form

$$\mathcal{S}(x,y) = \delta(x)\mathrm{rect}\frac{y}{vT} \tag{3.57}$$

with T being the integration time.

As shown by Goodman [22], the mean intensity of the integrated speckle pattern is equal to the mean intensity of the original speckle pattern. The variance of the intensity is

$$\begin{aligned} < I_0^2 > &= \tfrac{1}{S^2} \int\limits_{-\infty}^{\infty} \int\limits_{-\infty}^{\infty} \int\limits_{-\infty}^{\infty} \int\limits_{-\infty}^{\infty} \mathcal{S}(x_1,y_1)\mathcal{S}(x_2,y_2) \\ &\quad \times < I(x_1,y_1)I(x_2,y_2) > \mathrm{d}x_1\mathrm{d}x_2\mathrm{d}y_1\mathrm{d}y_2 \\ &= \tfrac{1}{S^2} \int\limits_{-\infty}^{\infty} \int\limits_{-\infty}^{\infty} R_{\mathcal{S}}(\Delta x,\Delta y)R_I(\Delta x,\Delta y)\mathrm{d}\Delta x\mathrm{d}\Delta y. \end{aligned} \tag{3.58}$$

The signal detected by a photomultiplier may also be written as the integral over the time interval $[0,T]$

$$I = \frac{1}{T}\int\limits_0^T |\mathbf{U}(\mathbf{r},t)|^2\mathrm{d}t. \tag{3.59}$$

Using this, Ohtsubo and Asakura [38] have shown that the probability density function of the integrated speckle intensity is obtained as the inverse Fourier transform of the characteristic function $< \exp(\mathrm{i}zI) >$, and given by

$$\begin{aligned} P_I &= \int\limits_{-\infty}^{\infty} < \exp(\mathrm{i}zI) > \exp(-\mathrm{rm}\ \mathrm{i}zI)dz \\ &= \tfrac{1}{2\pi} \int\limits_{-\infty}^{\infty} \frac{\exp(-\mathrm{i}zI)}{\prod\limits_{k=0}^{\infty}[1-\mathrm{i}z(\lambda/T)]}\mathrm{d}z. \end{aligned} \tag{3.60}$$

Assuming the amplitude $\mathbf{U}(\mathbf{r},t)$ obeys a complex Gaussian process, the last expression in [38] was transformed into

$$P_I = \exp\left(-\frac{v^2\tau^2}{w^2}\right) \tag{3.61}$$

where v is the velocity of the moving diffuse object, τ is the time correlation length of the speckle intensity variation, and w is the radius of the Gaussian beam focused on the diffuse object.

4. Speckle Field Generation and Recording

After the previous theoretical introduction to the speckle technique we are now going to describe the experimental methods used in speckle photography. The major components of an experimental system for speckle photography are

- light sources and optics,
- speckle field recorders,
- speckle field generators,
- automated data processing systems.

We will start with speckle generators. The description of lasers for speckle photography will be given shortly; in addition photographic and electronic recording systems will be described. One of the most important parts of speckle photography measurements, automated data processing systems, will be described separately in chapter 6.

4.1 Speckle generators

To produce a laser speckle pattern any rough surface can be successfully used, and a perfectly developed speckle pattern is not dependent on the rough surface parameters. Different methods of rough surface manufacturing can be used. Among them are ground glasses, photographic techniques, photoresist materials, and spraying a paint sealant. There is a great body of literature that covers the design of rough surfaces or diffusers with desired angular distribution of the scattered light and prediction of the intensity distribution of the light reflected from a given surface, see, e.g. [39-41]. Two main parameters of speckle field generators will be analysed here. They are the speckle statistics and the angular distribution of light behind the speckle generators.

The simplest speckle generator is a photographically recorded speckle field. Such a transparency may be made with different speckle sizes and its light distribution in the diffraction cone can be adjusted for a particular optical scheme. Very simple and reliable speckle generators are different types of ground and milk glasses. Gray describes the method of rough surface producing using photoresist materials [42] and Bonin and Kadar-Kallen used a spraying technique [43].

In the latter technique the random phase plates were simply made by spraying an acrylic paint sealant (colourless Krylon) on to a microscope slide. Different plates were held at different distances (25–100 cm) from the spray in order to produce different distributions of acrylic on the glass surface. For distances less than 30 cm a high density of large droplets was observed, so that the acrylic tended to flow before it completely dried. If the slide is held far away ($\simeq$100 cm) from the source, the droplets were observed to be smaller and well separated from each other when drying. It was determined in [43] that the acrylic droplet diameter varied from 10 to 200 μm. It was found experimentally that this simple diffuser had near-Gaussian statistics of the speckle pattern with speckle contrast equal to 0.92.

4.1.1 Angular distribution of scattered light

We will start from the relationship between the characteristics of a rough surface and the averaged angular distribution of light scattered from the surface. The latter value is described by the envelope of the normalized intensity distribution in a scattered speckled field [40]:

$$< I(\Omega) >= \frac{\Delta(< I >)}{\Delta\Omega I_0}. \tag{4.1}$$

Here $\Omega = \Omega(\theta, \phi)$ is a solid angle. For a plane wave impinging at normal incidence it is convenient to use a polar angle θ measured from the centre beam axis as shown in Fig. 4.1.

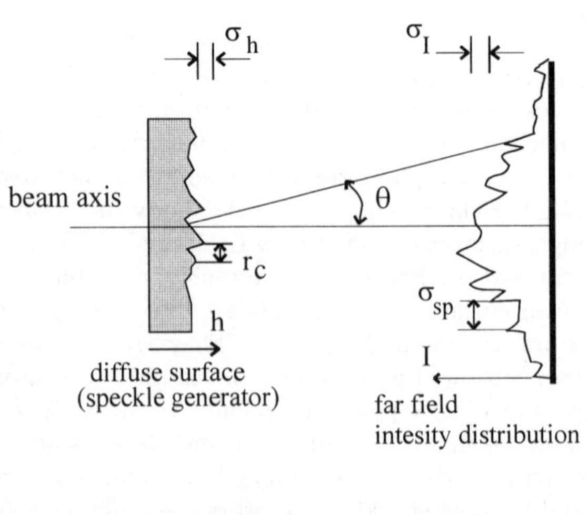

Fig. 4.1. Speckle field generation behind diffuser surface and some notation

The intensity pattern in the far field has an overall angular dependence in which the mean intensity decreases with increasing polar angle θ. This mean intensity envelope function is due to diffraction by the scattering elements of the diffuser. The light distribution in the diffraction zone is rather well defined only in the case when the diffuser for speckle generation is an idealized transparency with all scattering elements being perfectly circular, of equal area, and non-overlapping. Then the angular dependence of the mean intensity would be characterized by the well-known Airy function determined by the speckle size. However, due to the random shape and distribution of areas of the scattering elements, the actual envelope function is more complicated even for speckle generators made using specklegrams [43]. The same situation is found with the rough surface speckle generators. To describe this intensity distribution we need to describe firstly the rough surface itself. For a surface with random high fluctuations the same statistical description as for speckle fields can be applied.

Thus, it is possible to describe the rough surface by a height function $h(\xi, \eta)$ that represents the departure of the surface from it mean position.[1]

The normalized *autocorrelation function* of the surface is

$$R_{\mathrm{h}}(r) = \frac{< h(r_1)h(r_2) >}{\sigma_{\mathrm{h}}^2} \tag{4.2}$$

with σ_{h} being the *root mean square (rms) roughness*

$$\sigma_{\mathrm{h}} = \lim_{N\to\infty} \frac{1}{N}\left[\sum_{i=1}^{N} h_i^2\right]^{1/2} . \tag{4.3}$$

Here $< \ldots >$ indicates ensemble averaging. The function R_{h} is also called *a normalized autocovariance.*

Assuming the height variation to be random and isotropic, this function depends on the offset $r = [(\xi_2 - \xi_1)^2 + (\eta_2 - \eta_1)^2]^{\frac{1}{2}}$ between the points (r_1) and (r_2) but not on the direction of this offset. When the latter property holds, the envelope of the radiation pattern for normal incidence will be circularly symmetric.

Goodman [22] has use assumption that the normalized autocorrelation function of the surface heights is of Gaussian form:

$$R_{\mathrm{h}}(r) = exp[-(\frac{r}{r_{\mathrm{c}}})^2], \tag{4.4}$$

where r_{c} is correlation length of the surface roughness.

In contrast to this, Shirley and George [40] have used a representation of the autocorrelation function by a power series in r and analysed both conical

$$R_{\mathrm{h}}(r) = 1 - \frac{r}{r_{\mathrm{c}}} + \ldots, \tag{4.5}$$

and paraboloidal

---

[1] We will use here as usual the coordinates (ξ, η) for the rough surface (or object) plane and (x, y) for the observation (or image) plane.

$$R_{\mathrm{h}}(r) = 1 - (\frac{r}{r_{\mathrm{c}}})^2 + \ldots \tag{4.6}$$

forms of this dependence.

It can be noted that the latter is the expansion of the Gaussian autocorrelation function (Eq. 4.4) and the relation (4.5) is the expansion of an exponential autocorrelation function:

$$R_{\mathrm{h}}(r) = exp[-(\frac{r}{r_{\mathrm{c}}})], \tag{4.7}$$

The conical distribution is characteristic of a surface having discontinuities and high slopes. There is good agreement between the measured far field intensity for a ground glass speckle generator and the theory using the conical distribution. On the other hand, for etched glass speckle generators better agreement was obtained using the paraboloidal autocorrelation function, which is characteristic of smooth surfaces that are band-limited in spatial frequency [40].

The *power spectrum*, or *Wiener spectrum*, of the surface height variation gives the distribution of the spatial frequencies in the roughness and is given by the Fourier transform of the surface autocorrelation function [41]

$$W_{\mathrm{h}}(u,v) = \int_{-\infty}^{\infty}\int_{-\infty}^{\infty} R_{\mathrm{h}}(\Delta\xi,\Delta\eta)\exp[-2\pi\mathrm{i}(\Delta\xi\cdot u+\Delta\eta\cdot v)]\,\mathrm{d}(\Delta\xi)\mathrm{d}(\Delta\eta). \tag{4.8}$$

The relation between the high function and the complex amplitude of the light in the plane immediately to the right of the surface is an extremely complex one if it is developed with rigour [22]. Among the complicating factors are variations of the Fresnel reflection coefficient with slope, the effects of propagation from the point of reflection, multiple scattering, and shadowing. As a first approximation, the rms (variance) of the complex amplitude phase, Θ of the wave reflected or passing through the medium with a rough surface can be connected with the rms (variance) of h [22]:

$$\sigma_{\Theta}^2 = \sigma_{\mathrm{h}}^2 \cdot [k(1+\cos\beta)] \tag{4.9}$$

for the reflection, where β is the angle of illumination, and

$$\sigma_{\Theta}^2 = \sigma_{\mathrm{h}}^2 \cdot [k(n-1)] \tag{4.10}$$

for the wave passing through the medium with refractive index n at normal incidence.

The angular dependence of the intensity due to single and multiple speckle diffusers has been carefully studied and characterized [40]. The theory is particularly simple in the case of strong diffusers where the rms of the phase variation is large, i.e., $\Delta\Theta >> 1$. Thus, for a conical type of diffuser for normal incidence and observation in the plane of incidence the normalized radiant intensity was described as

$$< I(\theta) >= \cos\theta \cdot 2\pi(\frac{r_{\mathrm{C}}}{\lambda\sigma_{\Theta}^{2}})^{2}\{1 + (\frac{2\pi r_{\mathrm{C}}}{\lambda\sigma_{\Theta}^{2}})^{2} \cdot \sin^{2}\theta\}^{-\frac{3}{2}} \tag{4.11}$$

The experimental data for a ground glass distribution were best fitted over a dynamic range of 6 orders using the dependence (4.11) with parameter

$$\frac{r_{\mathrm{C}}}{\lambda\sigma_{\Theta}^{2}} = 1.4.$$

For a paraboloidal surface correlation function this expression has form

$$< I(\theta) >= \cos\theta \cdot \pi(\frac{r_{\mathrm{C}}}{\lambda\sigma_{\Theta}})^{2} \cdot \exp\{-(\frac{\pi r_{\mathrm{C}}}{\lambda\sigma_{\Theta}})^{2} \cdot \sin^{2}\theta\} \tag{4.12}$$

Analysis of these equations shows that the on-axis intensity of the conical diffuser is twice as great as that of the paraboloidal one for the same value of the appropriate shape parameter.

The etched glass radiation can be fitted over 3 order of magnitude by the equation (4.12). However, to fit the radiation pattern behind the etched glass over all angles, it is necessary to use a composite diffuser model containing two scales of roughness [40]. In this case the height can be represented as the sum of two normally distributed zero-mean components $h_1(r)$ and $h_2(r)$ that are uncorrelated with each other:

$$h(r) = h_1(r) + h_2(r) \tag{4.13}$$

and

$$\sigma_{\mathrm{h}}^{2} R_{\mathrm{h}}(r) = \sigma_{\mathrm{h}1}^{2} R_{\mathrm{h}1}(r) + \sigma_{\mathrm{h}2}^{2} R_{\mathrm{h}2}(r). \tag{4.14}$$

For etched glass with two characteristics of roughness different combinations of surface statistics can be used. In the frame of Shirley–George theory [40], there are three possible combinations of surface statistics for two-scale roughness, namely, conical–conical (CC), paraboloidal–paraboloidal (PP), and conical–paraboloidal (CP). For the (CC) combination, Eq. (4.11) will still hold with the effective shape parameter

$$\frac{r_{\mathrm{C}}}{\lambda\sigma_{\Theta}^{2}} = \left(\frac{\lambda\sigma_{\Theta 1}^{2}}{r_{\mathrm{C}1}} + \frac{\lambda\sigma_{\Theta 2}^{2}}{r_{\mathrm{C}2}}\right)^{-1}. \tag{4.15}$$

Likewise, for the (PP) combination, Eq. (4.12) will hold with effective shape parameter

$$\frac{r_{\mathrm{C}}}{\lambda\sigma_{\Theta}} = \left[\left(\frac{\lambda\sigma_{\Theta 1}}{r_{\mathrm{C}1}}\right)^{2} + \left(\frac{\lambda\sigma_{\Theta 2}}{r_{\mathrm{C}2}}\right)^{2}\right]^{-\frac{1}{2}}. \tag{4.16}$$

For the (CP) combination, the distribution can be found as the convolution of distributions described by Eq. (4.11) and Eq. (4.12).

As a result of the analysis made in [40], the next expression can be recommended for the etched glasses

$$
< I(\theta) >= \cos\theta \cdot \pi \exp(-\sigma_{\Theta 2}^{2}) \left[\left(\frac{r_{C1}}{\lambda \cdot \sigma_{\Theta 1}}\right)^{2} \exp\left\{ -\left(\frac{\pi r_{C1}}{\lambda \cdot \sigma_{\Theta 1}}\right)^{2} \cdot \sin^{2}\theta \right\} \right. \tag{4.17}
$$

$$
\left. +2\left(\frac{\sigma_{\Theta 2} \cdot r_{C2}}{\lambda}\right)^{2} \left\{ 1 + \left(\frac{2\pi r_{C2}}{\lambda}\right)^{2} \cdot \sin^{2}\theta \right\}^{-\frac{3}{2}} \right].
$$

The experimental data for etched glass were best fitted using the dependence (4.17) with parameters

$$
\frac{r_{C1}}{\lambda \sigma_{\Theta 1}} = 6, \qquad \frac{r_{C2}}{\lambda} = 0.75, \qquad \sigma_{\Theta 2} = 0.14.
$$

In Fig. 4.2 the scattering diagrams obtained in [40] for ground glass and etched glass diffusers are shown. In the same Fig. 4.2 the radiation distribution for an idealized Lambertian scattering surface

$$
< I(\theta) >= \frac{1}{\pi} \cos\theta \tag{4.18}
$$

is shown.

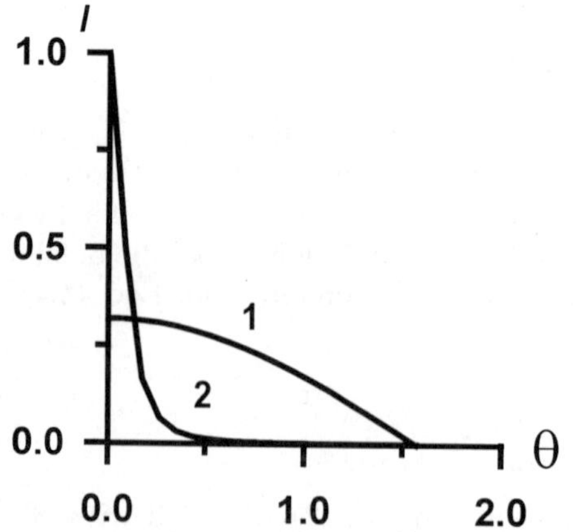

Fig. 4.2. Angular intensity distributions behind the Lambertian surface (1), and the ground glass diffuser (2)

For the acrylic paint sealant speckle generator Bonin and Kadar-Kallen [43] used the next expression for the angular intensity distribution obtained for a strong diffuser with a smooth surface in the small angle limit ($\theta << 1$):

$$< I(\theta) > \sim \exp\left[-(\frac{\theta}{b})^2\right], \quad (4.19)$$

where

$$b = \frac{\lambda\sigma_\theta}{r_C}. \quad (4.20)$$

The relation (4.19) corresponds to Eq. (4.12) for the paraboloidal surface correlation function in the small angle limit $\theta << 1$.

Experimental investigation of the acrylic paint sealant speckle generator [43] shows that the expression (4.19) describes the angular intensity distribution in the far field behind the generator and the best fit of the data using this expression yielded $b = 0.034$ for the particular conditions of the experiments.

4.1.2 Speckle statistics

As has already been mentioned, a perfectly developed speckle pattern is not dependent on the rough surface parameters. However in some cases we have a partially developed speckle pattern with speckle contrast less than unity. For such cases it is important to understand the conditions under which the statistical properties of the surface can affect the properties of the speckle pattern. In this respect we will briefly describe here the main correlations between a rough surface and the speckle field statistics.

As usual, let us restrict our consideration to monochromatic and fully polarized speckle patterns. We again use coordinates (ξ, η) for the plane adjacent to the scattering surface, and coordinates (x, y) for the observation plane.

For Gaussian surface statistics the complex coherence factor of the wave reflected from (or passing through) the rough surface takes the relatively simple form [22]

$$\mu_U(r) = \exp\{-\sigma_\Theta^2[1 - R_\mathrm{h}(r)]\} = \exp\{-\sigma_\Theta^2[1 - \exp -(\frac{r}{r_C})^2]\}. \quad (4.21)$$

Analysing this dependence we can see that for large σ_Θ^2, the width of $\mu_U(r)$ is much narrower than the width of $R_\mathrm{h}(r)$, indicating that the correlation area of the reflected (or transmitted) wave is far smaller than the correlation area of the surface [22]. Thus for large σ_Θ^2, the coherence properties of the reflected (or transmitted) light depend only on the behaviour of the surface correlation function near the origin. The normalized correlation function of the light amplitude for this case is

$$R_U(r) = \frac{\exp[-\sigma_\Theta^2(1 - R_\mathrm{h}(r))] - \exp(-\sigma_\Theta^2)}{1 - exp(-\sigma_\Theta^2)}. \quad (4.22)$$

It is convenient to introduce the coherence area of the light

$$S_c = 2\pi \int_0^\infty r \cdot |R_U(r)| \, dr. \tag{4.23}$$

The results of calculations of the dependence of this value on the surface roughness parameter again show that for large σ_Θ^2 the coherence area of the wave is much smaller than the correlation area of the surface itself.

The contrast of the speckle pattern obtained by reflection of coherent light from the rough surface was analysed by Fujii and Asakura [44]. For sufficiently large σ_Θ the speckle contrast saturates at its value of unity. This is the case when the speckle statistics are not dependent on the rough surface statistics.

For small σ_Θ, the contrast behaves asymptotically as

$$C \simeq \sigma_\Theta^2. \tag{4.24}$$

More detailed information about the influence of the rough surface statistics on the speckle statistics are contained in the specialized articles [22,39,41,45].

4.2 Light sources for speckle generation

The primary requirement of a light source in speckle photography experiments is that it provides sufficient coherent[2] light intensity during the exposure time adopted for the processes under study. Evidently, the choice of the light source will depend on the characteristic time of the processes studied and on the sensitivity of the recording system, which may be photographic or pure electronic.

For very slow processes a wide variety of continuous wave (CW) lasers was used and are also used today in speckle photography experiments. Thus for Bénard convection studies, Meynart [46,47], and Arroyo *et al.* [48-51] have used low-power CW lasers with mechanical shutters.

Vogel and Lauterborn have performed time resolved measurements of the temporal evolution of the flow fields, a bubble cavitation [52]. For this purpose the light source should be able to produce regular bursts of pulses of short duration and a high repetition rate. They have used a high power CW Ar^+ laser. The laser beam was chopped by an extra-cavity acousto-optic deflector controlled by a combination of two burst generators. Thus, pulse trains of variable length, pulse separation, and repetition rate can be generated. Each specklegram (PIV-frame in [52] was exposed with a sequence of five pulses with 1-μs duration and 3-μs distance between successive pulses. The pulse energy was 3 μJ.

---

[2] Systems using white light illumination for speckle photography will be discussed in the next chapter.

Using the same CW electric discharge laser medium excitation, it is possible to significantly increase the laser energy per pulse by introducing a similar optical modulator into the laser cavity and re-adjusting the laser into Q-switch operation mode. By doing this it is possible to increase the available laser energy approximately K times, where K is the ratio of the distance between successive pulses and the pulse duration. Such a laser was used for speckle photography in [53]. Essentially higher energy per pulse is achieved in pulsed lasers, among which the most widely used are ruby (Ru) and neodim (Nd:YAG) lasers. The first generate light in the visible range of the spectrum, $\lambda = 0.694\,\mu$m. This laser can be used both in a free generation mode and in Q-switched operation. In the latter mode a Pockels cell is inserted into the laser cavity and provides one, two, or a sequence of laser pulses during 1 ms with a time duration of each individual pulse of about 30 ns. In free generation operation a ruby laser provides one relatively long pulse with a time duration of about 1 ms. This operation mode is convenient for single (prolonged) exposure laser speckle photography (see chapter 5).

Using a Pockels cell it is possible to obtain two laser pulses with a time interval between the pulses varies from 1 μs to 0.5 ms. It is possible to obtain an arbitrary time interval between two subsequent laser pulses using two synchronized ruby lasers. Such a configuration was used, e.g. by Blinkov *et al.* [54] to provide simultaneous multiangular test section illumination in multidirectional speckle photography for tomographic reconstruction of a 3D flow field.

A ruby laser can be supplied with amplifiers. Usually, a laser with one rod generates about of 0.1–1 J of coherent energy, whereas with amplifiers this energy could be increased to 5–10 J.[3]

There is another available high-power pulsed laser in the visible region of the spectrum – a neodim (Nd:YAG) laser. Pumped by light emitted from a flashlamp, a neodim rod generates energy in the near infrared (IR) spectral region, $\lambda = 1.064\,\mu$m. The second harmonic generator (KDP crystal) provides 0.532 μm green light. Usually, one rod laser with a Pockels cell in a Q-switch operation mode generates coherent energy of about 0.2–0.5 J with a time duration of the laser pulse of 5–6 ns. It is possible to generate two subsequent pulses or a sequence of pulses using by controlling a Pockels cell with appropriate electrical signal. Often, two synchronized neodim lasers are operated to provide two subsequent laser pulses with an arbitrary time interval between the pulses [55].

To extend the spectrum of the neodim laser it is possible to use third- and fourth-harmonic generators which provide 355 nm and 266 nm UV light, respectively.

Laser beam divergence and beam pointing stability are also important for speckle photography applications. For neodim lasers these parameters are about of 0.5 mrad (full angle, measured at 90% of energy profile) and about

[3] There are ruby laser with laser energy of more than 1 kJ.

100 μrad. Time jitter ± 0.5 ns and energy stability of $\pm 2\%$ could be easily achieved.

A very high energy both in CW and pulsed modes can be obtained by using IR CO_2 lasers [56,57]. The average power in CW mode of about 100 W is typical for lasers of this type, but specific infrared recording materials are needed for the IR spectral region; see the next section.

Among the high repetition rate operation mode lasers is the Cu-vapour laser, pumped by an electric discharge. The laser of this type provides coherent radiation with wavelength 504 nm and an average power about of 20–50 W. At a frequency of 10 KHz the laser provides a continuous sequence of laser pulses with duration of about 30–50 ns and a time interval between two subsequent pulses of 100 μs. The energy of each pulse is about of 2–5 mJ. Such lasers were used for speckle photography with additional choppers in publications [55,58].

Going back to the low-power lasers, we need to mention diode lasers [59]. These lasers are now used in *electronic speckle pattern interferometry*, ESPI, the recording part of which will be discussed in the Section 4.4. Such lasers are very cheap and reliable, but need very sensitive sensors. For instance, diode laser used in [59] has an output power which could be varied between 0 and 10 mW at a wavelength of 0.845 μm.

Some parameters of lasers used in speckle photography are summarized in Table 4.1.

4.3 Photographic recording

Photographic recording is widely used in speckle photography to detect and store optical information for subsequent analysis. A unique feature of the photographic recording technique is its great memory capacity. In speckle photography, like in holography, one photographic plate with typical size 9 by 12 cm can contain information about more than 10^6 images[4] each of which need several Mb of computer memory if we use electronic memory to store it. Of course, there are special mathematical methods to compress images by tens or even hundred of times to save computer memory, but even taking this into account, photographic recording is still out of the competition in terms of memory capacity.

The second salient feature of photographic recording is it high spatial resolution. Till now modern electronic recording systems have had a spatial resolution of about 50–100 lines/mm with limited size of recording media, usually about 2 by 2 cm. This value of electronic recording media resolution is determined by a typical pixel size of about 20 by 20 μm in CCD arrays. As can be seen from Table 4.2, the resolution power of the films is essentially higher.

[4] We assume here that the interrogation diameter is about 0.1 mm.

Table 4.1. Characteristics of lasers for obtaining specklegrams

Type and short description	Wavelength, in μm	Power or energy per pulse	Remarks or reference
Ar^+, CW mode	0.5145	2 W	[60]
Ar^+, CW mode with a Bragg cell modulator	0.5145	5–7 W	[61,62]
Ar^+, CW with outside acousto-optical deflector	0.5145	3.6 W 3 μJ per pulse	frequency up to 1 MHz, [52]
Ar^+, CW with intra-cavity acousto-optical deflector	0.5145	1 W	[53]
Ru, pulsed laser free-generation	0.694	1–10 J	[21]
Ru, pulsed laser Q-switch operation	0.694	0.2–10 J	[21,63,64]
YAG-Nd, pulsed laser Q-switch operation	0.532	0.2 J	[65]
Cu-vapour laser repetition rate operation	0.504	20 W 2 mJ at 10 KHz	[55]
Infrared CO_2 CW and pulsed lasers	10.6 10.6	20–100 W 1–1000 J	[57] [66]

Despite these two main advantages, photographic recording has many disadvantages.

One of the first is non-linearity of the photographic processes. It is usually assumed, that there is a linear part in the dependence of $D(\log E)$, where D is the photographic density of the developed film and E is the 'exposure', that is the energy density per unit area of the light, adsorbed by the film emulsion.

4.3.1 Photographic emulsions

The photographic density D is determined through the film transmittance:

$$D = \log(\frac{1}{|T|^2}) \tag{4.25}$$

where T is the amplitude transmittance and $|T|^2$ is the intensity transmittance [10].

Table 4.2. Characteristics of films for recording specklegrams

Type	Resolution, line/mm	Exposure for $D = 1$, μJ/cm^2	Remarks
Agfa 8E75	5 000	10	$\lambda = 632.8$ nm
Agfa 8E56	5 000	15	$\lambda = 476$ nm
		25	$\lambda = 521$ nm
Agfa 10E56	2 800	1.4	$\lambda = 476$ nm
		1.9	$\lambda = 521$ nm
Agfa 10E75	2 800	2	$\lambda = 632.8$ nm
Agfa 14C70	1 500	0.3	$\lambda = 700$ nm
Agfa 14C75	1 500	0.3	$\lambda = 700$ nm
Kodak 649F	2 000	70	$\lambda = 632.8$ nm
		500	$\lambda = 694.3$ nm
Kodak 125	1 250	2	$\lambda = 441.6$ nm
		5	$\lambda = 514.5$ nm
Kodak 131	1 250	0.5-0.8	$\lambda = 632.8$ nm
Kodak 120-02	1 250	33	$\lambda = 694.3$ nm
Kodak TP 2415 (ASA 125)	320	0.15	$\lambda = 694.3$ nm
Kodak Royal-XP (ASA 1 250)	100		400–620 nm
Kodak Tri-X (ASA 400)	80	0.02	$\lambda = 694.3$ nm
Polaroid Type 55 (ASA 50)	150	0.1	$\lambda = 514.5$ nm, $D = 0.5$
Polaroid Type 57 (ASA 3000)	20	0.001	$\lambda = 514.5$ nm, $D = 0.5$

Photographic films are often characterized by the slope γ of this linear portion of the sensitometric curve:

$$D = D_0 + \gamma_{\mathrm{D}} \cdot \log(E). \tag{4.26}$$

In speckle photography, like in holography, amplitude transmittance is of more direct use than density, so sensitometric data are often displayed as real amplitude transmittance versus exposure[5]

$$T = T_0 - \gamma_T \cdot E. \tag{4.27}$$

The value of the exposure in these relations can be expressed as the product of the irradiance and the exposure time:

[5] γ_{T} can be expressed through γ_{D}.

$$E = \int_{(\Delta t)} I(t)\,\mathrm{d}t. \tag{4.28}$$

The constants γ in Eqs. (4.26) and (4.27) are dependent on the development processes. If quantitative or semi-quantitative data about the photographic density of the film are used, the development processes must be controlled very precisely.

In speckle photography, like in holography, the bleaching process can be used to convert amplitude transparency to phase transparency, see [10], p. 52–53. This procedure can significantly improve the contrast of Young's interferometric fringes, obtained during specklegram treatment.

The obvious disadvantage of photographic recording is the necessity of using 'wet' development processes. These processes need time and in most cases the photographic plate must be moved from the position where the exposure was made.

It should also be noted that the sensitivity of photographic materials is rather low at the wavelengths of the widely used ruby and helium-neon lasers (see Table 4.2). This is not true for panchromatic films with extended and maximum sensitivity in the red.

Another parameter to be accounted for is the film grain [61]. When illuminating the film negative to produce Young's fringes, the film's unexposed grains can introduce amplitude and/or phase changes into the wavefront of the analysing laser beam and thus create additional noise. This noise source can be reduced by producing a positive copy by contact printing, on a very high-resolution fine grain film [62,67]. It was shown in [68,69] that for low-resolution films the use of the liquid gate will improve the accuracy of photographic recording and decrease SNR.

A time-consuming step in speckle photography is the photographic processing of the specklegram. The processing time can be reduced several orders of magnitudes if thermoplastic (TP) recording materials are used. TP recording takes intermediate place between the photographic and electronic methods of speckle pattern recording both in the operation time and in the spatial resolution. Compared with photographic film, the spatial resolution of TP materials is lower, but much higher than for electronic recording by CCD cameras, whereas the operation time for CCD cameras is much shorter compared to TP cameras.

A metal-thermoplastic (MTP) material has already been used both for speckle photography in the IR region of the spectrum [57] and a photo-thermoplastic (PTP) material for particle image velocimetry (PIV) [70]. The MTP camera for speckle photography was developed at the Research Institute of Applied Physical Problems of the Belarus State University [57]. The carrier has a three-layer structure and consists of a sublayer (milar film about 100 μm thick), a thin metal film and a thermoplastic layer. The thin metal film absorbing IR radiation non selectively is used for recording and heating the thermoplastic layer about 1 μm thick. Polyepoxycarbazole serves as the thermoplastic.

The carrier is heated to a temperature close to that at which there occurs the 'frosty' deformation of the surface of the thermoplastic charged to a critical potential to record IR radiation. The thermoplastic is charged immediately before it is exposed, for example, with the help of a coronator. During exposing the prepared carrier, the thin metal film absorbs the recording radiation energy and locally heats the thermoplastic layer according to the spatial distribution of the recording radiation energy. Local heating of thermoplastic sections results in the onset of 'frosty' deformation on its surface due to electrostatic forces. In this case the deformation amplitude is proportional to the heating temperature of the thermoplastic section. The energy distribution of IR radiation thus recorded is fixed due to the spreading of the thermal relief developed in the thermoplastic layer and due to carrier cooling. The obtained image is then visualized by laser beam illumination of the exposed carrier. Note that both the sensitivity of the MTP carrier (about 10 $\mathrm{mJ/cm^2}$) and the spatial resolution ($\sim 200\,\mathrm{mm}^{-1}$) substantially exceed the corresponding characteristics of the recording media that are traditionally used in IR holography such as, for example, polymethylmetacrylate, cellulose triacetate, chalcogenide glassy semiconductors, *etc.* [71].

The MTP carrier was fastened in a unit (camera) incorporating a coronator, thermal heater and thermal sensor, and was connected with an electronic control block. The control block ensured a supply of a 5–10 kW voltage to the coronator and thermal stabilization of the carrier temperature within about 0.1 ^{o}C. Good-quality Young's fringes were obtained using this unit both for λ =1.06 and 10.6 μm.

As an example of the PTP, we will describe the properties of the roll film RF/3 (Rothech, Germany), which has been used in PIV [70]. This film consists of a transparent plastic substrate, a photoconductive layer and an insulating thermoplastic layer. Both layers have a thickness of about 1 μm. The substrate side of the film is located on a gold coated glass plate that is connected to ground potential. Before exposure, a corona discharge in air generates a uniform positive charge distribution on the surface of the thermoplastic layer, resulting in an electric field between this surface and the glass plate. The exposure with a recorded light intensity distribution produces free charge carriers in the photoconductor, whose density is proportional to the light intensity. The electric field forces the positive charge carriers to move in the direction to the boundary between the photoconductor and the substrate, while the negative charge carriers move in the opposite direction. Subsequently, the ohmic heating of the gold layer on the glass plate softens the thermoplast, so that it may be deformed by the Coulomb forces due to the charge distribution. Thus a thickness variation of the thermoplastic layer is produced, which is proportional to the recorded light intensity distributions.

The spatial resolution of PTP materials strongly depends on the thickness of the thermoplastic layer and the development processes [72] and reaches values of about 400–600 mm^{-1}. It was found that the required exposures for PIV records on PTP materials is about 0.2 J/m^2 both for frequency doubled Nd:YAG lasers and for ruby lasers [70]. This is the same order of magnitude as for Agfa Holotest films 8E56/8E75.

Therefore, the experimental investigations [57,70] show that MTP and PTP materials may be used for specklegram recording. The visibility of Young's fringes achievable with currently available materials is sufficient for automated fringes processing. The use of MTP carriers allows one to extend the applicability of speckle photography methods in the IR region of the spectrum to the fast recording of speckle fields. This opens up the possibilities of speckle photography of IR media, and substantially increases the sensitivity of speckle photography to electron concentrations in plasma flows.

4.4 Electronic recording

Specklegram processing is practically always performed electronically, so an electronic reading system is an essential part of any speckle photography installation, see chapter 6. In this section we will concentrated on the description of the first step of speckle photography only, obtaining the specklegram.

The attractiveness of electronic recording is explained by the possibility of direct input of primary experimental data into the computer without a 'wet' processing. The major difficulty with such recordings is associated with the low spatial and temporal resolution of present electronic system.

The early 1980s showed rapid progress in the development of solid-state imaging devices [73]. Nowadays, a wide variety of electro-optical devices including vidicons, plumbicons, tivicons, and many solid-state imagers are used for light intensity distribution recording [74]. A charge coupled device (CCD), sometimes also called a charge coupled detector, is the most widely used. CCD cameras are also used for easy and fast image acquisition in speckle photography. Using these devices, an image can be stored directly in digitized form on a computer and then analysed numerically.

A typical CCD is a metal-oxide-semiconductor that is composed of a two-dimensional rectangular array of closely spaced independent sites, called pixels, where a photon-induced charge is stored [60]. The principal advantages of the CCD are its high quantum efficiency and low noise. The spatial resolution of the CCD recording is determined by the pixel dimension, which is now about 10 by 10 μm^2 and by the pixel array size. The number of pixels in an array is varied from typical values of 384×576 [60] to 5000×7000 [75]. This is equivalent to a photographic film with resolution about of 50 line pairs per millimetre. The minimum energy required for a good recording on the CCD is about $0.7 \cdot 10^{-4}$ $\mu J/cm^2$ which is at least three orders of magnitude less than even for the best high-resolution films.

The use of electronic recording in holography and speckle interferometry began in the early 1979s in several scientific groups [76,77] and is reviewed in many publications [4,74,78-80].

The main components of an electronic recording system are a CCD camera with electronic unit and interfaces, a computer with frame memory, A/D and D/A converters, and, optionally, a real time digital video recorder, see Fig. 4.3.

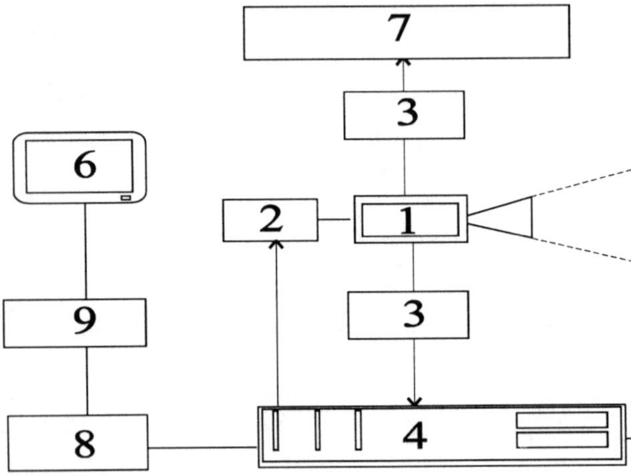

Fig. 4.3. The main components of electronic speckle recordings. 1, camera head with electronic unit; 2, camera controller; 3, interfaces; 4, host computer; 5, computer monitor; 6, video monitor; 7, real time digital video recorder; 8, frame memory; 9, D/A convertor

Image acquisition systems can be broadly classified as *real time* or *static* [74]. Usually, the real time system has a standard video refresh rate equal to 25 or 30 frame/s. Higher refresh rates are generally difficult to achieve. When the video rate is insufficient, one of the possible alternatives is to use a multiple CCD camera unit [81-84]; another is to use cameras that can be made to acquire two images only with a short separation and a long subsequent period until the next two images can be acquired [85,86].

Table 4.3 contains the number of 8-bit bytes necessary to store a single frame in computer memory as a function of spatial resolution, expressed as the number of pixels per side of the square sensor (image format).

Another characteristic that has to be considered is *the signal-to-noise ratio* (SNR). The dynamic range of a sensor is usually given as the ratio between the maximum peak signal and the rms noise. A typical value for CCDs at room temperature is about 500:1 [74]. The noise is generally expressed as the current generated by the sensor when no light impinges on it. This *dark*

Table 4.3. Memory requirements as a function of CCD camera resolution (image format) and A/D conversion resolution (image depth)

Image format, pixels/side	Image depth, bits/pixel	Total memory, Kbytes	Image depth, bits/pixel	Total memory, Kbytes
128	8	16	24	48
256	8	64	24	192
512	8	256	24	768
1 024	8	1 024	24	3 072
2 048	8	4 096	24	12 288

current shows a very strong dependence on temperature. Cooling of the sensor may raise the dynamic range up to 10 000:1 [74].

The image, acquired by a CCD camera, has to be digitized by an A/D convertor. The fundamental characteristics of an A/D convertor are the resolution of the conversion (image depth) and the speed at which the video signal must be sampled. A typical resolution of 8-bits corresponds to a resolution of 256 grey levels. The number of reliable bits that can be extracted from a digitization system is a function of the SNR of the system.

Videocameras avoid the post-processing procedures necessary for photographic film and make image acquisition much faster. Since video cameras record time sequences the use of separate images of single-exposed particle images removes the directional ambiguity in the recovered data and also allows the displacement measurement of partially overlapping particles images.

Let us describe some details of existing installations performing electronic specklegram recording. Willert and Gharib have used electronic recording for PIV measurements [87]. A CW 5 W Ar^+ laser was shuttered at a 30 Hz rate with a pulse duration of 4 ms for periodic flow illumination. Shuttering was necessary because, similar to photographic film, the imaging array of the camera integrates over the time period between frame capture (1/30 s).

The image acquisition system consisted of a Hitachi C1H CCD colour camera of which only the green signal was sampled at 8 bits/pixel by a Vidicon image digitizer and processor. Digitized images have an active area of 512 pixels horizontally by 480 pixels vertically, which, in the case of the Willert and Gharib experiments, corresponds to a field a view of approximately 8.3 cm by 7.7 cm, or a scaling factor of 6.13 pixels/mm. The flow images were written to a real-time digital video-disk at a framing rate of 30 Hz. This disk was capable of recording 750 images, or 15s worth, and can then be downloaded and archived on a magnetic tape as conventional binary data files. The technique using digitally recorded video images rather than photographic transparencies and applied to velocity measurements by particle recordings was named *digital particle image velocimetry*, DPIV [87]. Since the use of separate images of single-exposed particle images in DPIV removes the directional ambiguity in the recovered data by preserving the order of the images, this also allows the displacement measurement of partially overlapping particle images.

Merzkirch *et al.* have employed this technique for measuring the velocity distribution of the natural convective flow [88]. In contrast to the Willert and Gharib recording scheme, who stored the experimental information on a real time disk in digital form, Merzkirch *et al.* used analogue recording of the frames on video type with subsequent frame digitizing. In this experiment the CCD camera had a format of 756 $\times$ 581 pixels and the video frequency was 25 frames per second. After the experiment, the signal from the video type was given to the frame grabber of a PC. The frame grabber was capable storing the information of 24 video frames in digital form with a resolution of 8 bits (256 grey values) per pixel. Further processing of the digitized particle image patterns by applying the cross-correlation algorithm (see chapter 6) was performed with the PC. A monitor was adjusted to observe on-line the motion of the tracer particles in the flow.

In [83] the video cameras were equipped with a Sony XC-77 RR-CE CCD chip with a size of 756$\times$581 pixels. The analogue video signal is transmitted to a frame grabber installed in a personal computer. The frame grabber board digitizes the video signal and stores it into its own fast memory. Via this memory access is given to the image data. The image then has a size of 512$\times$512 pixels with a colour depth of eight bits, which means that the grey values range from 0 (black) to 255(white).

As compared to holography, speckle photography needs at least one order of magnitude lower spatial resolution of the recording media. It is still a high enough resolution of order $\sim$ 200–1000 line/mm, which is well satisfied for photographic emulsions and near to the limit of the spatial resolution of modern TV systems. In the next chapter we will describe a number of successful applications of TV recording systems for speckle photography, based on a direct correlation analysis of the varied speckle patterns. In this section we can emphasize, that requirements for spatial resolution of the recording media in the case of speckle interferometry are even lower than in the case of speckle photography, and the possibility of using electronic recording for this case is one of the salient feature of speckle interferometry. This has the obvious advantage that it converts the visual information into an electrical signal, which can readily be processed electronically.

5. Basic Optical Schemes for Obtaining Specklegrams

In this chapter a general description of some common ideas widely used in speckle photography will be given. The main attention will be paid to obtaining specklegrams, while the treatment of specklegrams will be described in the next chapter.

5.1 Speckle photography

Let us start from the simplest schemes of speckle photography, sometimes called single-beam speckle interferometry [89]. The starting point of utilizing this technique in metrological applications was Burch and Tokarski's publication [1]. They have shown that the Fourier transform of any small part of a double exposed specklegram is equivalent to the Fourier transform of two orifices, the diameters of which are equal to the average speckle size and the spacing between which is equal to the speckle shift at the analysed point of the specklegram.

For *in-plane* surface displacement measurements speckle photography was introduced by Archbold *et al.* [90,91] and later explored and reviewed by many researchers [2,6,92-95].

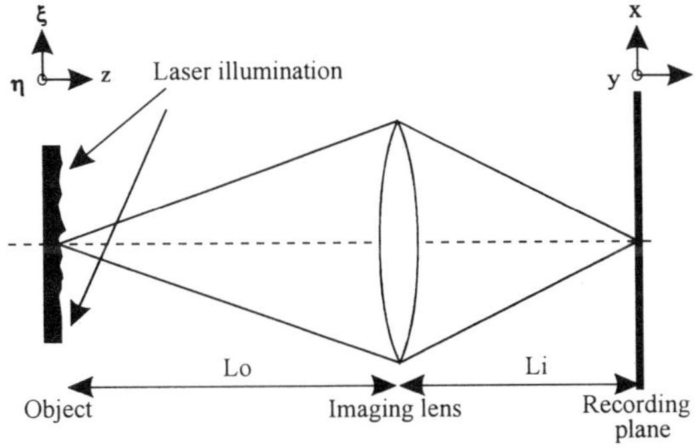

Fig. 5.1. Speckle photography of in-plane surface displacement

The simplest form of speckle photography involves the straightforward double exposure recording of a laser-illuminated object in two different positions or states of stress [96]. Subsequent probing of the processed specklegram with an unexpanded laser beam produces a set of interferometric (Young's) fringes with angular orientation being perpendicular to the direction of displacement, and whose spacing is inversely proportional to the in-plane surface displacement. A schematic diagram of the method is illustrated in Fig. 5.1. The object to be studied is illuminated by a coherent beam from any angle, and the speckle pattern produced is imaged by the camera or observed near the scattering surface. As can be seen, all that is required is a laser for illumination and a camera with high-resolution film. By doubly exposing the film in the image plane with the specimen being deformed between the exposures, a fringes pattern depicting in-plane displacement due to deformation can be extracted from the processed photographic plate using a Fourier filtering technique. Both the component of the in-plane displacement as well as the sensitivity can be selected during the filtering processes by varying the distance of the filtering aperture from the optical axis; see the next chapter. At the specklegram recording stage, the sensitivity can be altered by adjusting *the magnification ratio*, $M = \Delta_{\xi,\eta}/\Delta_{x.y}$, where $\Delta_{\xi,\eta}$ is the local surface displacement and $\Delta_{x.y}$ is the image of this displacement at the recording plane. Another version of the single beam method is to use an imaging lens with double or quadruple apertures [89,92]; see Section 5.1.1.

The single beam method has an advantage over both holographic interferometry and dual beam speckle interferometry, in that it does not require special vibrational isolation. Further, the interpretation of the fringes does not depend upon the illumination direction, thus allowing diverging beam illumination and thereby eliminating the size limitation of the object.

It seems appropriate to compare the single beam method with the Moiré method, as both are sensitive to in-plane displacement [97]. While the Moiré method requires printing a grating on the object surface, the single beam method utilizes the naturally generated speckles as the grating. The sensitivity of the single beam method is about an order of magnitude higher than that for the Moiré method. The inherently poor fringes quality is a limitation of the speckle method.

5.1.1 Speckle photography of in-plane surface displacement (deformation)

By taking in turn a sequence of moving speckle pattern images into a photograph, called *a specklegram*, it is possible to determine the average speckle displacement at each small area of the specklegram. The speckle motion can be related to the movement or deformation of a rough surface, which generates a speckle pattern. Thus, the movement or deformation of a rough surface can be quantitatively measured by illumination of this surface by coherent light and recording of the produced speckle pattern. To demonstrate this, let us denote by $I(x, y)$ or $I(\mathbf{r})$ the intensity distribution of the speckle field at a point $\mathbf{r} = \mathbf{i} \cdot x + \mathbf{j} \cdot y$ in the recording plane.

Double exposure operation. We suppose that two exposures of this speckle pattern on the photographic plate were made and between these exposures the speckles were shifted, for instance, by the distance $\boldsymbol{\Delta} = \mathbf{i} \cdot \Delta_x + \mathbf{j} \cdot \Delta_y$. Therefore, during the second exposure, the speckle pattern in the plane of the photographic plate is given by $I(\mathbf{r} + \boldsymbol{\Delta})$. It is convenient to present these fields as results of the convolution with δ-functions:

$$I(\mathbf{r}) = I(\mathbf{r}) \otimes \delta(\mathbf{r}) \tag{5.1}$$
$$I(\mathbf{r} + \boldsymbol{\Delta}) = I(\mathbf{r}) \otimes \delta(\mathbf{r} + \boldsymbol{\Delta}).$$

Under the assumption linearity of the photographic processes and omitting a constant term, the amplitude transmittance of a double exposed specklegram can be expressed as:

$$T(\mathbf{r}) = b \cdot D(\mathbf{r}) = T_1 + T_2 \tag{5.2}$$

where $D(x, y) = [I(\mathbf{r}) + I(\mathbf{r} + \boldsymbol{\Delta}] \cdot \Delta t$, and Δt is the exposure time.

Upon illumination of such a specklegram with a small-diameter collimated probe beam, the amplitude distribution in the diffraction halo region (far field) is given by the Fourier transform of the specklegram transmittance function. Denoting this amplitude by $U_{\mathrm{F}}(\mathbf{r}_1)$, where $\mathbf{r}_1$ is the coordinate in the Fourier plane, $\mathbf{r}_1 = \mathbf{i} \cdot x_1 + \mathbf{j} \cdot y_1$, one can obtain

$$U_{\mathrm{F}}(\mathbf{r}_1) = \mathcal{F}\{D(\mathbf{r})\} = U_1 + U_2, \tag{5.3}$$

where U_1 and U_2 are the amplitude distribution at the specklegram plane related directly to the speckle patterns during the first and second exposures. Correspondingly, the intensity distribution at the Fourier plane is

$$I_{\mathrm{F}}(\mathbf{r}_1) = |U_{\mathrm{F}}(\mathbf{r}_1)|^2 = U_1 \cdot U_1^* + U_2 \cdot U_2^* + U_1 \cdot U_2^* + U_2 \cdot U_1^* \tag{5.4}$$

where

$$U_1(\mathbf{r}_1) = \int\int_{(\sigma)} T_1(\mathbf{r}) \exp(\mathrm{i}2\pi \frac{\mathbf{r}_1 \cdot \mathbf{r}}{\lambda f})\, \mathrm{d}x\mathrm{d}y \tag{5.5}$$

and

$$U_2(\mathbf{r}_1) = \int\int_{(\sigma)} T_2(\mathbf{r} + \boldsymbol{\Delta}) \exp(\mathrm{i}2\pi \frac{\mathbf{r}_1 \cdot (\mathbf{r} + \boldsymbol{\Delta})}{\lambda f})\, \mathrm{d}x\mathrm{d}y. \tag{5.6}$$

Integration here is performed through the so-called interrogation area, e.g. the area on the specklegram illuminated by the laser beam.

The crossed terms in Eq. (5.4) represent Young's interferometric fringes enveloped by a diffraction halo of the specklegram. As a first approximation, this distribution may be written as

$$I_{\mathrm{F}}(\mathbf{r}_1) = I_{\mathrm{h}}(\mathbf{r}_1) \cdot \left\{1 + \mathrm{Re}[\exp(\mathrm{i}2\pi \frac{\mathbf{r}_1 \cdot \boldsymbol{\Delta}}{\lambda f})]\right\} \tag{5.7}$$

where Re denotes the real part of a complex function.

This relation can be rewritten as

$$I_{\mathrm{F}}(\mathbf{r}_1) = I_{\mathrm{h}}(\mathbf{r}_1) \cdot \cos^2 \left[\frac{k(\mathbf{r}_1 \cdot \boldsymbol{\Delta})}{2L}\right] \tag{5.8}$$

It is seen that the diffraction halo of a single exposure specklegram is now modulated by a series of uniformly spaced cosine square fringes whose spacing is inversely proportional to $|\boldsymbol{\Delta}|$, and whose direction is normal to $\boldsymbol{\Delta}$.

The dark fringes are observed when

$$\mathbf{r}_1 \cdot \boldsymbol{\Delta} = (n + \frac{1}{2})\lambda L, \quad n = 0, \pm 1, \pm 2, \ldots \tag{5.9}$$

and bright fringes when

$$\mathbf{r}_1 \cdot \boldsymbol{\Delta} = n\lambda L, \quad n = 0, \pm 1, \pm 2, \ldots \tag{5.10}$$

Equations (5.9) and (5.10) can be combined as

$$|\boldsymbol{\Delta}| = \frac{\lambda L}{d_{\mathrm{f}}}, \tag{5.11}$$

where d_{f} is the fringe spacing and L is the distance from specklegram to the observation plane.

Multiple-exposure technique. Making successively several exposures of a speckle pattern on to one photographic plate with equal displacement between exposures it is possible to increase the fringes contrast in the treatment stage of the specklegram. In fact, let us make a series of successive exposures with equal exposures times Δt_n and equal translations between each exposure, $\boldsymbol{\Delta}$. Then the transparency of the specklegram will be changed with each exposure by the value

$$T(\mathbf{r}) = I_0(\mathbf{r}) \otimes \delta(\mathbf{r} + n\boldsymbol{\Delta}) \cdot \Delta t_n \tag{5.12}$$

and its transparency after $N + 1$ exposures will be given by

$$T(\mathbf{r}) = \sum_{n=0}^{N} T_n(\mathbf{r}). \tag{5.13}$$

Under illumination of the developed specklegram by a thin laser beam as described in the previous section, in the diffraction halo one can observe the intensity distribution, proportional to the Fourier transform of the function $T(\mathbf{r})$ [3]:

$$I(\mathbf{r}_1) = I_{\mathrm{h}}(\mathbf{r}_1) \cdot \left[\frac{\sin\left\{ \frac{(N+1)\pi \mathbf{r}_1 \cdot \boldsymbol{\Delta}}{\lambda L} \right\}}{\sin(\frac{\pi \mathbf{r}_1 \cdot \boldsymbol{\Delta}}{\lambda L})} \right]^2 . \tag{5.14}$$

This distribution is the same as for a grating with a period equal to Δ and the number of bars equal to $(N+1)$. Between each of the two principal maxima in this distribution there are $N-1$ secondary maxima. As shown by Françon, [3], the secondary maxima can be suppressed if the exposure times $\tau_1, \tau_2, \ldots$, etc., are proportional to the binomial coefficients C_N^P, see Table 5.1. The spacing between maxima in the distribution (5.14) is the same as for the double exposure specklegram, namely

$$d_{\mathrm{f}} = \frac{\lambda L}{|\boldsymbol{\Delta}|} \tag{5.15}$$

and the sought value of the displacement may be determined using expression (5.11).

Table 5.1. The required ratio of exposure times to suppress the secondary maxima in distribution (5.14)

Number of exposures	Modulation factor	Ratio of exposure times
3	$\cos^4\left(\frac{\pi \Delta r_1}{\lambda L}\right)$	1:2:1
5	$\cos^8\left(\frac{\pi \Delta r_1}{\lambda L}\right)$	1:4:6:4:1

So, using simple measurements of the fringes spacing it is possible to determine quantitatively the value of the speckle shift, $\boldsymbol{\Delta}$, at each small *interrogation* area of the specklegram.

Single exposure speckle photography. The double (and multiple) exposure operation assumes the very short exposures compared to the characteristic flow time, during which a recorded speckle pattern is 'frozen'. The same optical schemes can be used for recording a speckle pattern during prolonged exposures (compared to the speckle motion time). This provides *time averaged* speckle pattern recording and can be used for the measurement and analysis of a time averaged surface deformation or flow field.

This operation has been used for the surface vibration analysis since publication of Eliasson and Mottier [98]. At the same time, a similar approach, single exposure operation using speckle interferometry, was used for vibration analysis in a number of publications [99-101], and is reviewed in [102]. The schemes of speckle interferometry will be described in Section 5.1.2.

Both in *speckle photography* and in *speckle interferometry* schemes, in regions where speckles are moving during exposure time, the *contrast* of the speckle pattern is changed. In *reference wave speckle interferometry,* this is due to different statistics of the intensity distribution in different specklegram regions. Ek and Molin [100] showed that the speckle contrast C varies with the vibration amplitude A according to the equation

$$C = [1 + 2\alpha \mathcal{J}_0(4\pi A/\lambda)]^{1/2}/(1+\alpha), \tag{5.16}$$

where α is the ratio of the reference beam intensity to that of the object beam, and $\mathcal{J}_0$ is the Bessel function of order zero and the first kind.

Similarly, in *speckle photography* with prolonged exposure time if an object surface is moving in and out with an amplitude on the order of one-quarter of a wavelength or more, its speckle pattern will become blurred out and the *contrast* of speckles will become very poor [102]. However, for stationary areas of the surface, the high speckle contrast will remain. Thus, the nodal regions of vibration can be detected and mapped [98,103]. Using photographic speckle pattern recording, it is difficult to distinguish between specklegram areas with different speckle contrast. A further spatial filtering of the specklegram enhances the contrast variation; see chapter 6.

For flow measurements the term *single exposure speckle photography* was introduced by Fercher and Briers under velocity mapping of human retina blood flux [104]. As for rough surface movement, the speckle pattern of the specklegram of a living human retina in an area where flow is occurring will be blurred to an extent which will depend on the velocity of flow and on the exposure time. The same phenomenon was observed when a speckle pattern was formed under illumination of living plants by laser light [105]. Going back to a random surface velocity distribution, like blood flux in a human retina, and introducing the speckle intensity correlation time, $\tau_C \sim 1/\text{velocity}$, Fercher and Briers [104] derived an expression for *the single exposure* speckle pattern contrast

$$C = [(\tau_C/2T)\{1 - \exp(-2T/\tau_C)\}]^{1/2} \tag{5.17}$$

where T is the exposure time, and the speckle statistics were obtained for a Lorentzian distribution of velocities with normalized intensity autocorrelation function $R_I(\tau)$ being

$$R_I(\tau) = \exp(-2\tau/\tau_C). \tag{5.18}$$

For a Gaussian velocities distribution the speckle contrast is [106]

$$C = [\frac{\sqrt{\pi}\tau_C}{2T} \cdot \text{erf}(T/\tau_C)]^{1/2}, \tag{5.19}$$

whereas for the constant velocity case it is [106]

$$C = [\frac{1}{T} \cdot \int_0^T \{\frac{2\mathcal{J}_1(\frac{\pi v \tau}{F\lambda})}{\frac{\pi v \tau}{F\lambda}}\}^2 \, \mathrm{d}\tau]^{1/2}, \tag{5.20}$$

where v is the required velocity, F if the f-number of the imaging lens, and $\mathcal{J}_1$ is the Bessel function of the first order.

Contrast enhancement may be carried out by means of a straightforward high-pass filtering technique [107] or by using digital recording with subsequent computer-aided processing [108].

In [109] single exposure speckle photography of a rotating disk was used for angular velocity determination by diffraction halo analysis; see below. The same approach, *single exposure speckle photography* using *light sheet illumination*, was used in [110] for measuring the velocity profiles in fluid flows. In both cases the expression (5.24) was used for the diffraction halo analysis.

For line-of-sight optical methods this technique was adopted for turbulent flow diagnostics in publications [111-114] where it was referred to as *single (prolonged) exposure speckle photography* (SPESP). The optical scheme for the SPESP is the same as for line-of-sight double exposure speckle photography, see below. The speckle movement during the single prolonged exposure here occurred due to integration over the optical path density fluctuations. Small speckles are produced here during a short exposure. On the other hand, the prolonged exposure time used and the turbulent speckle motion results in a relatively large speckle size. The form of speckles, magnified with a microscope, is shown in Fig. 5.2 for four different positions in a single exposure specklegram [111].

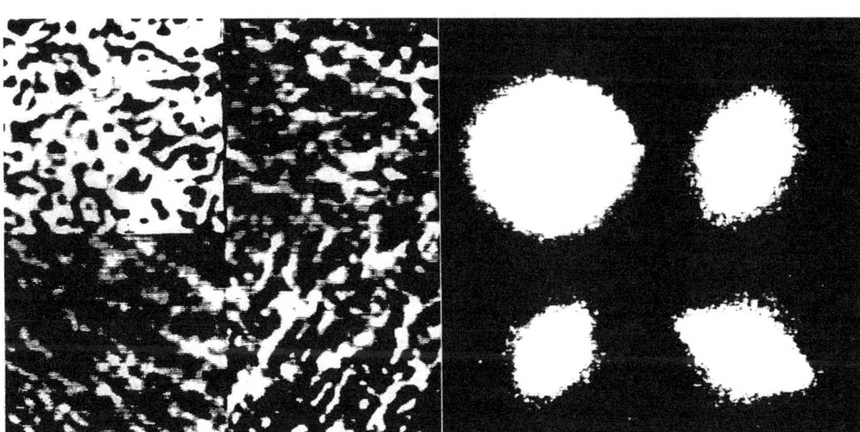

Fig. 5.2. Speckle form at four different positions of a single exposure specklegram(left). Magnification 30x and corresponding cross-section of diffraction halo for these four positions (right)

The diffraction halo analysis is performed by interrogation of the specklegram with a thin laser beam. The halo from the single exposures is not modulated by Young's fringes. The halo cross-section is circular if the turbulent fluctuations in a plane normal to the direction of light propagation are evenly distributed, as in isotropic turbulence. With anisotropic turbulence, the speckles become elongated in the preferential direction of turbulence, and the halo-cross section is elliptical with the longer axis being normal to the direction of the elongated speckles.

To describe the intensity distribution in the diffraction zone behind a single exposure specklegram, let us a consider continuous recording of a moving speckle pattern on a film during prolonged exposure time. For the object velocity at the object plane $\mathbf{v}$, the velocity of the speckle pattern at the recording plane is $M\mathbf{v}$, where M is the magnification of the imaging system. In the pulse duration (width) the object moves by

$$\boldsymbol{\Delta} = M\mathbf{v}\tau \tag{5.21}$$

with τ being the illumination pulse duration. For simplicity we assume a displacement along one direction only, when $\mathbf{v} = v$ and $\boldsymbol{\Delta} = \Delta$. The amplitude transmittance of the photographic plate owing to this exposure can be expressed as

$$T = t_0 - \beta\{D(x,y) \otimes \mathrm{rect}(\Delta), \tag{5.22}$$

where t_0 is the bias term, β is the film constant, and $D(x,y)$ is the intensity distribution in *the reference speckle pattern.* In this case it is a speckle pattern obtained at the same optical scheme without the flow.

In the pointwise filtering method, the Fourier transform (FT) of the amplitude transmittance occurs, and for the amplitude at the observation plane we have

$$U = \mathcal{F}\{T\} = \mathcal{F}\{t_0\} - \beta\mathcal{F}\{D\}(f_x, f_y) \cdot \mathrm{sinc}\left(\frac{\Delta f_x}{\lambda L}\right) \tag{5.23}$$

with L being the distance from the specklegram to the observation plane.

Omitting the dc term, the intensity distribution at the observation plane is

$$\begin{aligned} I = |U|^2 &= \beta|\overline{D}|^2 \cdot \mathrm{sinc}^2\left(\tfrac{\Delta f_x}{\lambda L}\right) \\ &= I_{\mathrm{h1}} \cdot \mathrm{sinc}^2\left(\tfrac{\Delta f_x}{\lambda L}\right), \end{aligned} \tag{5.24}$$

where

$$\overline{D} = \beta\mathcal{F}\{D\}(f_x, f_y).$$

For double exposure speckle photography this intensity distribution may be treated as a diffraction halo, which is modulated by the signal due to *a transit* of the speckle pattern to the distance $\delta = vt_0$, where t_0 is the time interval between two subsequent pulses [115]

$$\begin{aligned} I &= \beta|\overline{D}|^2 \cdot \mathrm{sinc}^2\left(\frac{\Delta f_x}{\lambda L}\right) \cdot \cos^2\left(\frac{\delta f_x}{\lambda L}\right) \\ &= I_{\mathrm{h2}} \cdot \cos^2\left(\frac{\delta f_x}{\lambda L}\right). \end{aligned} \tag{5.25}$$

The SPESP technique can be used both in frame and in streak recording mode [116,117]. As in streak mode for high-speed photography, a slit is installed to record the time history of a single exposure speckle pattern along a selected line in the flow studied (Fig. 5.3).

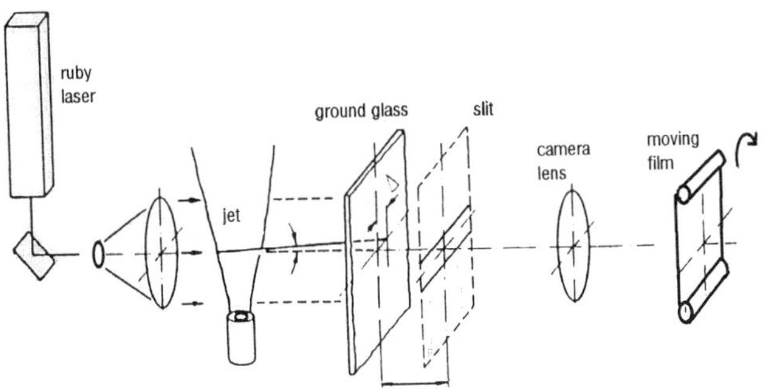

Fig. 5.3. Experimental installation for the SPESP in streak mode

Individual speckles in this mode will form streaks with different thickness on the moving film, similar to interferometric fringes. The thickness of these fringes is typically about 1–10 μm, of the same order as the speckle size for the SPESP. This thickness at a chosen film position will be determined by the intensity of the density (or temperature) fluctuations at a chosen instant of time.

The results of the application of the single exposure technique for turbulence studies will be described in chapter 9.

Double-aperture speckle photography. In double-aperture operation mode (see Fig. 5.4) the interferometric (Moiré) fringes can be observed on the processed specklegram directly and the further spatial filtering of the specklegram can be employed to enhance the fringes contrast only. In this mode a speckle pattern is formed in the image plane with large speckles relative to full-aperture speckle photography. Due to the interference of two overlapping speckle patterns from two apertures, reference interferometric fringes are formed. The spacing of these fringes is

$$d_{\text{ref}} = \frac{\lambda L_2}{D} \tag{5.26}$$

where D is the distance between the apertures.

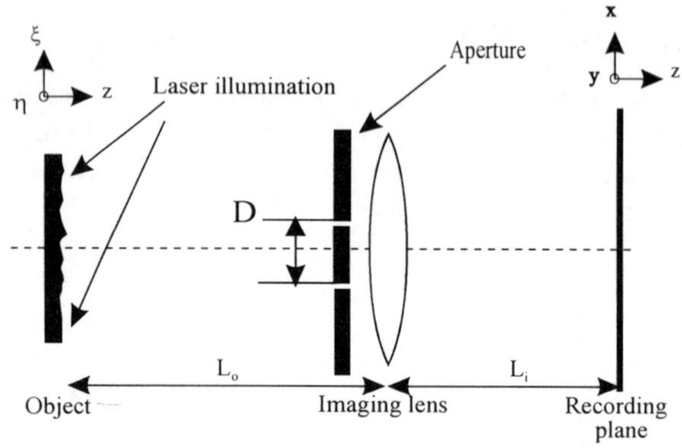

Fig. 5.4. Schematic diagram of double-aperture speckle photography

In double-exposure operation, a small local displacement of the object, occurring between the two exposures, in a direction parallel to a line joining the two apertures, will generate Moiré fringes, which are a measure of the displacement. By incorporating four apertures in a cross-geometry, both x and y components of motion can be evaluated [93,118].

This two-aperture recording technique also permits the observation of displacement fringes in real time if a single-exposure speckle photograph of the non-stationary process is recorded, and precisely replaced in the set-up prior to the displacement of the object [96].

However, the technique requires considerably increased exposures due to the severely limited aperture area. Furthermore, aperturing such that only areas at the edge of the lens are utilized reduces its image-forming qualities [96].

White-light speckle photography. *White light speckle photography* was introduced into speckle metrology for measuring deformation, strain and shape of a rough surface [119,120]. The method is essentially the same as *the laser speckle method*, however *a white light source* is used in the recording stage instead of a laser. The method is very close to the *Moiré technique*, but based on random structures analysis. Such random structures could either be created artificially or may be present naturally on the surface of the test object. These random structures are similar to the laser speckle patterns, but may be recorded using white light illumination. Further *white light specklegram* processing is accomplished using the same coherent methods as for *laser speckle photography*; see chapter 6.

Many methods of artificial speckle pattern creation are reviewed by Asundi, [121]. The simplest is to first paint the surface white and then spray it with a fine mist of black paint to give the speckled appearance. A retroreflective paint consisting of small glass beads suspended in liquids is widely used in white light speckle pattern formation. Any specklegram, obtained under laser light illumination, can be further used as a speckle pattern under white light illumination. For many materials like concrete, machined aluminium, wood, etc., the natural roughness of the object can be used for 'quasi-speckle pattern' formation.

The recording of white light speckle patterns requires high-quality optics and precise focusing [121]. Unlike laser speckles, the speckle size in the white light speckle method depends on the aberrations of the imaging lens.

The attractiveness of the white light speckle method is the possibility of applying the technique to large objects and objects outside the laboratory. The technique is widely used now in experimental mechanics for the measurement of surface deformation and movement of small and large objects [122-126]. White light speckle photography has also been applied to deformation measurement in the interior of objects [127] and for fluid mechanical measurements [128-132]. Such measurements will be discussed later in chapter 8.

A modern extension of white light speckle photography for fluid velocity measurements is velocimetry, based on the flow image cross-correlation analysis [133-135]. In this technique, speckles are small fragments of a flow image obtained at different instants of time by using flow visualization methods. This technique will also be described in chapter 8.

The further improvements. Let us consider some present limitations of speckle photography.

The first is the range of measured values. Using point-by-point optical specklegram treatment (see chapter 6), to obtain Young's interferometric fringes the smallest lateral displacement of a speckle pattern between exposures must be greater than the smallest speckle size, i.e.

$$|\boldsymbol{\Delta}| \geq 1.22 M \lambda F \tag{5.27}$$

with M being the optical magnification, and F being the f-number of the recording lens. Thus the smallest displacements are compared with the wavelength of the light used. Nevertheless, there have been successful attempts at measuring displacements of even less than the speckle size.

One of the simplest ways to do that is to introduce, as proposed by Vikram and Vedam [136], an additional image shift of the order of the speckle size between two exposures. In this case it is possible to analyse Young's interference fringes even in the regions where the initial speckle shift between two exposures is less than the speckle size. So, using lower values of the F number for the imaging lens, speckle photography with an additional image shift between exposures can be utilized to measure submicrometre displacements of a speckle pattern.

The sub-speckle-size displacement measurements can be performed also by speckle photography with digital auto- or cross-correlation analysis of the images of a speckle pattern. By digital speckle pattern recording with subsequent mathematical treatment, sub-pixel accuracy can be achieved under speckle displacement measurements [137].

However, the minimal pixel size in this case is essentially greater than the minimal possible speckle size under photographic speckle pattern recording.

Concerning the biggest values of displacements which can be measured by this technique, speckle photography has limitations due to the finite correlation length of the speckle pattern. Till now we have assumed that the contrast of Young's interferometric fringes in double exposure speckle photography is equal to unity. In reality, many factors can conspire to decrease this contrast. One of the most important among them is the so-called decorrelation effect. For the case of speckle photography this effect was studied both theoretically and experimentally [138,139]. Usually, good contrast fringes can be obtained at the values of displacements $\Delta \leq 20\sigma_{\mathrm{sp}}$, and the value of σ_{sp} can be easily varied by appropriate choice of the F number of the imaging lens.

It is necessary to take into account that interferometric fringes in the Fourier plane are modulated by speckle noise. Several reduction systems, based on averaging principles, have been designed to overcome the adverse effects of this speckle noise [140,141]. The distribution of intensity in the diffraction halo has an effect on the Young fringes positions. This effect will be analysed in the next chapter. Here we will mention one of the possibilities to eliminate this effect, proposed by Joenathan and Sirohi, [142]. This proposition is based on the possibility of obtaining a rather uniform distribution of intensity in diffraction halo regions by using specially designed apertures in the stage of specklegram recording. So, a rectangular aperture with two open sections of width $\eta_1 a$ and $(1-\eta_2)a$, placed in front of an imaging lens, produce a near to linear light distribution when $\eta_1 = 1/2, \eta_2 = 3/4$ or $\eta_1 = 3/4, \eta_2 = 1/2$.

5.1.2 Speckle photography of a 'transparent' medium

Line-of-sight speckle photography. In all fluid mechanical applications of photographic recording of a flow pattern it is necessary *to visualize* the desired parameter variation in *a transparent* medium using one or another optical technique [12]. In the case of speckle photography such a parameter is the density variation, which deflects a light beam passing through the flow studied. This beam deflection can be easily transformed into the speckle shift in the speckle photography method. Two possible methods of this transformation are shown in Figs 5.5 and 5.6.

In the first method the object under study is placed in front of the speckle generator, and in the second method, behind it. For both configurations the speckle shift in the image plane is related to the deflection angles of the light passing through the object, ϵ, by the equation

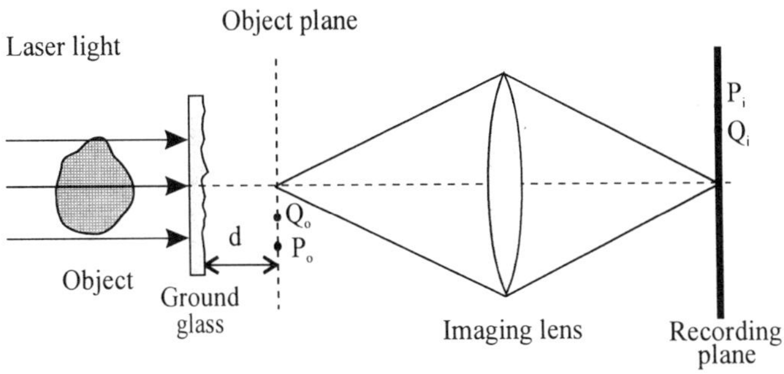

Fig. 5.5. Speckle photography of transparent object with the object interposed between the light source and the speckle generator

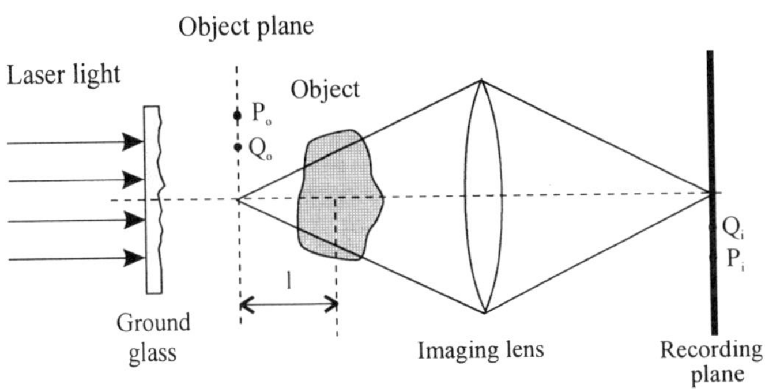

Fig. 5.6. Speckle photography of transparent object with the speckle generator interposed between the light source and the object

$$\Delta = M \cdot d \cdot \epsilon \tag{5.28}$$

where M is the optical magnification of the lens and d is the 'defocusing' distance.

Based on these considerations double exposure scheme photography was proposed for fluid mechanical applications by Debrus *et al.* [16], by Köpf [14,15] and by Mallick and Roblin [17]. They have proposed using one exposure of the speckle pattern without the flow field (this exposure can be named *the reference*) and another with the flow field introduced into the optical path of the light forming a speckle pattern. Using an analogy with holography, the second exposure can be classified as *the object exposure*. Due to refraction on the density gradients, the light beams during the object exposure are deflected and the speckle shifts are produced at each region of a specklegram corresponding to different flow regions. Ground glass was used there as a speckle field generator.

In both schemes an *objective* speckle at given points $\mathrm{P_o}, \mathrm{Q_o}$ in the *object plane* is formed by the interference of the light scattered by the speckle generator. At conjugate points $\mathrm{P_i}, \mathrm{Q_i}$ in the *image plane* a *subjective* speckle is formed by the interference of all rays that arise from points $\mathrm{P_o}, \mathrm{Q_o}$ and reach points $\mathrm{P_i}, \mathrm{Q_i}$ via the imaging lens.

At the surface of the photographic plate a speckle field is formed due to random interference of coherent laser radiation scattered by the ground glass. The amplitude of this field can be expressed through the Fresnel–Kirchhoff formula in the Fresnel approximation (see Eq. 2.29):

$$\begin{aligned} U_1(x,y) &= \frac{\mathrm{i}\cdot\exp(\mathrm{i}kd)}{\lambda d}\cdot\int_{-\infty}^{\infty}\int_{-\infty}^{\infty} U_{\mathrm{D}}(\xi,\eta) \\ &\times\exp\left(-\mathrm{i}k\cdot\left[\frac{(x-\xi)^2+(y-\eta)^2}{2d}\right]\right)\mathrm{d}\xi\mathrm{d}\eta. \end{aligned} \tag{5.29}$$

where $U_{\mathrm{D}}(\xi,\eta)$ is the amplitude distribution on the ground glass (diffuser). If we place the transparent phase object before the diffuser, the incident laser beam is deviated by the angle ϵ due to the presence of density (temperature) gradients in the object. It was shown elsewhere that this deviation of the incident beam is equivalent to the translation of the light distribution in the plane of the photographic plate [138]. Indeed,

$$\begin{aligned} U_2(x,y) &= \frac{\mathrm{i}\cdot\exp(\mathrm{i}kd)}{\lambda d}\cdot\int_{-\infty}^{\infty}\int_{-\infty}^{\infty} U_{\mathrm{D}}(\xi,\eta'-\epsilon d)\exp(\mathrm{i}k\epsilon\eta) \\ &\times\exp\left(-\mathrm{i}k\cdot\left[\frac{(x-\xi)^2+(y-\eta)^2}{2d}\right]\right)\mathrm{d}\xi\mathrm{d}\eta, \end{aligned} \tag{5.30}$$

where $\eta' = \eta + \epsilon d$. Taking into account that

$$\epsilon\eta + \frac{(y-\eta)^2}{2d} = \frac{\epsilon^2 d}{2} + 2\epsilon y + \frac{[y-\eta']^2}{2d}, \tag{5.31}$$

the amplitude distribution on the photographic plate in the presence of the object will be

$$U_2(x,y) = C\cdot U_1(x, y-\epsilon d), \tag{5.32}$$

where $C = \exp[\frac{\epsilon^2 d}{2} + 2\epsilon y]$.

So, regardless of the coefficient C, which is near unity, deviation of the beam through an angle ϵ is equivalent to a translation of the ground glass through ϵd.

If we make the first exposure of the speckle field on the photographic plate in the absence of the object, the plate receives an intensity distribution $I_1(x,y) \sim |U_1(x,y)|^2$. The second exposure should be made in the presence of the object under study. At this time the plate receives an intensity distribution $I_2(x,y) \sim |U_2(x,y)|^2$. The amplitude transmittance of the processed plate in the case of linear photographic processes can be determined through these intensities and the exposure time, τ:

$$T(x,y) = a - b\tau(I_1 + I_2). \tag{5.33}$$

Burch and Tokarski [1] have shown both experimentally and theoretically that when two identical speckle patterns are recorded on a photographic plate which has been laterally translated by a short distance between exposures, the diffraction halo generated by the processed plate will consist of a pattern of parallel straight fringes similar to Young's interferometric fringes. Indeed, considering the amplitude distribution in the diffraction halo region behind the illuminated specklegram containing the exposures of two speckle patterns, $U_3(w,v)$, as a Fourier transform of the amplitude transmission function $T(x,y)$ of the processed specklegram, we can write

$$U_3(w,v) = \exp(\frac{\mathrm{i}Lw^2v^2}{2k}) \int_{-\infty}^{\infty}\int_{-\infty}^{\infty} T(x,y)\exp(-\mathrm{i}wx - \mathrm{i}vy)\,\mathrm{d}x\mathrm{d}y. \tag{5.34}$$

Let us suppress for brevity the x coordinate [1], then

$$\begin{aligned} U_3(v) &= \exp(\tfrac{\mathrm{i}Lv^2}{2k}) \textstyle\int_{-\infty}^{\infty} T(y)\exp(-\mathrm{i}vy)\,\mathrm{d}y \\ &= \exp(\tfrac{\mathrm{i}Lv^2}{2k})\mathcal{F}\{T(y)\} \\ &= \exp(\tfrac{\mathrm{i}Lv^2}{2k})\mathcal{F}\{a + b\tau[|U_1(y)|^2 + |U_1(y+\epsilon d)|^2]\} \\ &= \exp(\tfrac{\mathrm{i}Lv^2}{2k})(a\delta(v) - b\tau\mathcal{F}\{|U_1(y)|^2\})(1+\exp(\mathrm{i}v\epsilon d)). \end{aligned} \tag{5.35}$$

It is convenient to have this expression rewritten in the form proposed in [1]:

$$I_3(v) = C_3|I^0|^2|S(v)|^2, \tag{5.36}$$

where $C = b^2\tau^2$,

$$I^0(v) = \mathcal{F}\{|U^1(y)|^2\}, \tag{5.37}$$

and

$$|S(v)|^2 = \cos^2\frac{v\epsilon d}{2}, \tag{5.38}$$

is the normalized interference term. The intensity distribution $I^0(v)$ corresponds to a diffraction halo of a single exposure specklegram, in which no displacement is involved. This halo is modulated by the term described in Eq. (5.38) and measures the distance between interferometric fringes, d_f, so we can determine the value of the speckle shift between two exposures:

$$\epsilon d = \frac{L\lambda}{d_\mathrm{f}} \tag{5.39}$$

With regard to the investigation of transparent phase objects, such measurements give quantitative information about density gradients in the object (see Section 7.1.1).

In experiments [14,15,17] the flow field was inserted between the ground glass and the recording lens (see Fig. 5.4). Later, this optical scheme was used in a number of quantitative measurements in fluid mechanical applications e.g. [18,21,143,144].

An inconvenience of this scheme is connected with the fact that the distance d is not less than one-half of the object size. For an extended object this distance is too big. It should be noted that the area of spatial averaging of data at the recording stage in this scheme is proportional to this distance d. This area could be roughly estimated as

$$s_{\mathrm{av}} = \frac{\pi}{4}\left[\frac{dD}{2f}\right]^2 \tag{5.40}$$

with D being the lens diameter and f its focal length.

The scheme shown in Fig. 5.3 is free from this disadvantage because by using such a scheme it is possible to decrease this 'defocusing' distance d as much as we want. The trade-off between sensitivity which is increased with d (Eq. 5.28) and the spatial averaging described by Eq. (5.40) determines the choice of d. This optical scheme was originally put forward by Debrus *et al.* [16] and modified by Wernekinck and Merzkirch for very extended objects [145]. This modified scheme is shown in Fig. 5.5 and is now widely used in speckle photography experiments [146-152]. As compared to the optical scheme shown in Fig. 5.5, an additional lens was placed in the modified scheme which allows one to direct a light beam passing through an extended object to the same area of the ground glass as a 'reference' beam, e.g. the same beam during *a reference exposure* when the object under study is absent. This modification allows one to decrease the decorrelation effect due to the use of the same area of the ground glass for generation of a speckle pattern both for *the object* and for *the reference* exposures. The sensitivity of this scheme is described by the equation

$$\Delta = d \cdot \epsilon \cdot \frac{M_2}{M_1} \tag{5.41}$$

with M_1 and M_2 being the magnifications of the first and the second lens, and can be easily varied by the variation of the distance d. The *defocusing area* for this scheme is also defined by the same equation as for the scheme shown in Fig. 5.5, namely Eq. (5.40).

Detailed comparison of these two measurement schemes made in [151] show that the contrast of the Young's interferometric fringes in the scheme shown in Fig. 5.5 is usually better, so the precision of the measurements using this optical scheme is high.

Light sheet illumination. As early as 1976 Chiang [153] suggested the use of a sheet of coherent illumination as a means of obtaining 3D interior surface displacement vectors by using the illumination of a transparent model in three

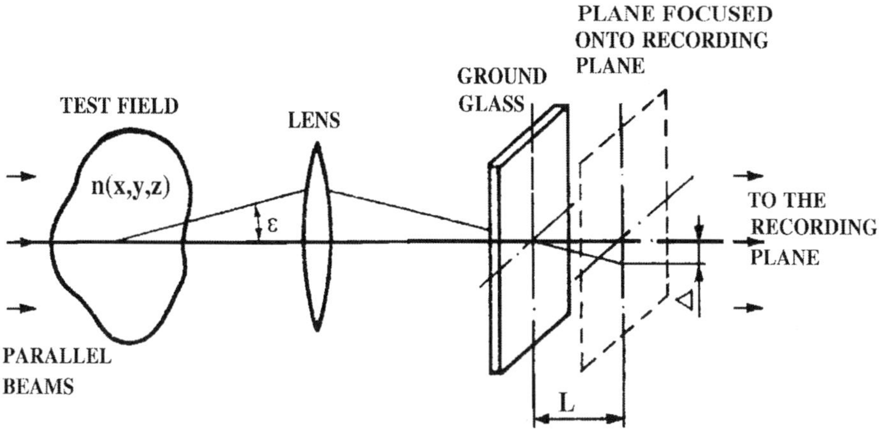

Fig. 5.7. Optical set-up for speckle photography of an extended transparent object (from Wernekinck and Merzkirch [145])

orthogonal directions through the point of interest and speckle photography of the scattered light from the direction normal to the incident beam.

In the same year, 1976, Barker and Fourney were the first to use the laser sheet and speckle photography method to obtain data about the deformation field of the interior part of a transparent object [154]. They obtained high-quality specklegrams using light scattered from a thin sheet of coherent light passing through a Plexiglas model and showed that these specklegrams could be analysed to yield the interior surface in-plane displacement vectors in the illuminated plane.

In the following year this technique was extended to the measurement of the velocity field of a fluid flow seeded with tracer particles and illuminated by a laser sheet [155-159].

For this technique Dudderar and Simpkins have used the term scattered light speckle (SLS) metrology, see [160].

Based on the measurements of the simultaneous displacements of images of many marked fluid particles, this technique was later named laser speckle velocimetry (LSV), particle image velocimetry (PIV), pulsed laser velocimetry (PLV), or particle image displacement velocimetry (PDIV). The term *particle image velocimetry*, PIV, is now the most frequently used.

The wide variety of optical schemes based on the use of the speckle technique for velocity measurements today may be schematically divided into two principal classes. The first class includes the schemes based on *laser sheet* illumination and *a static speckle pattern* recording allows one to reconstruct an *instantaneous* two-dimensional velocity map (Fig. 5.8). The second one is based on methods where space-time correlation functions are determined by using *dynamic speckle pattern* recording. Both techniques will be described in more detail later in chapter 8.

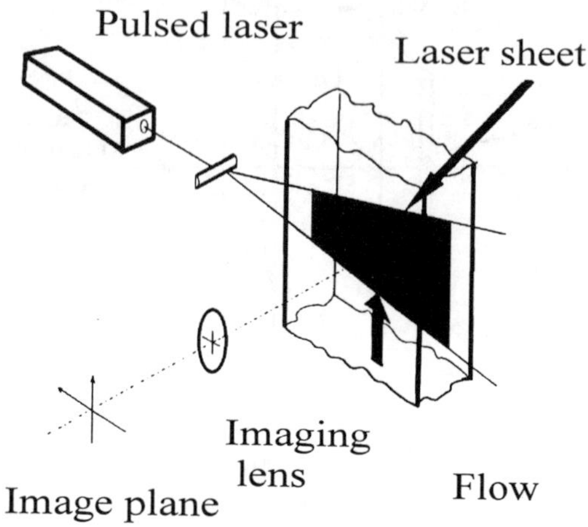

Fig. 5.8. Light sheet speckle photography of velocity field

5.2 Speckle interferometry

Introduced by Leendertz [161], *speckle interferomerty* is widely used in *speckle metrology* [23,78,89,93,96,162-165].

A great variety techniques have been reported on the utilization of *speckle interferometry.* As remarked by Erf [96] the variety of speckle interferometer configurations seems almost directly proportional to the number of experimentalists working in the field. There are various classifications of speckle interferometry and many definitions, when the technique belongs to *speckle photography* or *speckle interferometry.*

Thus Ennos [23] grouped the speckle techniques under two heads: (1) those based on the positional shift of the speckles (speckle photography) and (2) those based on the irradiance changes of the speckles (speckle interferometry).

Stetson [93] describes speckle photography as the technique where there are well-correlated regions in the two speckle patterns and the fringes are result from translation of two speckle patterns. When the fringes are derived from fluctuations in the correlation of the two speckle patterns, Stetson calls it speckle interferometry [96].

Hung [89] classified speckle technique into two categories: speckle interferometry and speckle-shearing interferometry. While the former measures surface displacement, the latter determines the displacement derivatives. In turn, Hung divided speckle interferometry into two versions: the dual beam method and the single beam method, or speckle photography, which is described above.

Both speckle interferometry and speckle shear interferometry can be performed using electronic detection, and the technique is broadly known as *electronic speckle pattern interferometry* (ESPI) [165].

A significant feature of speckle interferometry is that its sensitivity is higher than in speckle photography and comparable to that obtained in *holographic interferometry* [4].

Let us consider some examples of speckle interferometers.

5.2.1 Speckle interferometry with a reference beam

The dual-beam method was first introduced by Leendertz [161] and later investigated by many authors [163,166-168].

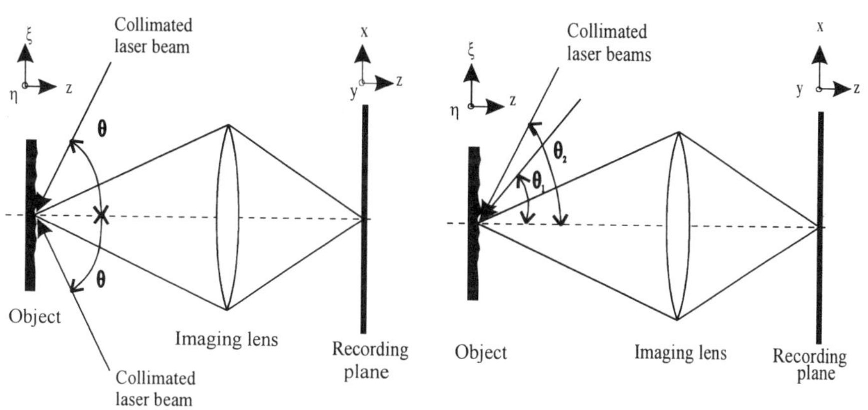

Fig. 5.9. Schematic diagram of the two dual-beam speckle interferometers: on the left, for measuring in-plane displacement, and on the right, for measuring out-of-plane displacement

Two optical arrangements of the dual beam method used to measure *in-plane* (on the left) and *out-of-plane* displacements are shown in Fig. 5.9. The object to be measured is illuminated by two collimated coherent beams. For *in-plane* measurements these beams are equally inclined at the angle θ to the yz plane as shown, and the object is imaged by the camera lens. With this arrangement the two speckle wavefronts, one due to each beam, impinge on the photographic plate. For each object point there are thus two rays that interfere at the image plane. When summed over the entire object the result is a speckle pattern that is recorded by the photographic film.

When the object is deformed, the displacement of the points causes a relative phase change between these two speckle wavefronts. The optical path changes due to the displacement along the viewing direction (*out-of-plane* displacement), and the *in-plane* displacement in the y-direction are the same for both beams, and will not give rise to any change in the relative phase difference of the two independent speckle patterns. However, an x-direction displacement, Δ_x, will cause a phase change $\Delta\Theta$,

$$\Delta\Theta(x,y) = \frac{4\pi}{\lambda}\sin\theta\Delta_x(x,y). \tag{5.42}$$

For *out-of-plane* measurements the specimen to be studied is illuminated simultaneously by two collimated beams inclined at angles θ_1 and θ_2, respectively, to the viewing direction. In this case, one beam serves as a reference for the other and the relative phase change between the two speckle wavefronts is given by

$$\Delta\Theta(x,y) = \frac{2\pi}{\lambda}\left\{(\cos\theta_1 - \cos\theta_2)\cdot\Delta_z(x,y) + (\sin\theta_1 - \sin\theta_2)\cdot\Delta_y(x,y)\right\}. \tag{5.43}$$

Thus this interferometer alignment is sensitive to *in-plane* displacement along the y-axis and *out-of-plane* displacement along the z- axis. The major advantage of this version is that the sensitivity can be reduced when θ_1 approaches θ_2 so that it is suitable for determining relatively large *out-of-plane* displacements.

Comparing the resulting speckle patterns, one before, and another after the surface displacement, there will be reduced correlation between the two speckle patterns in the regions where

$$\Delta\Theta = (2n+1)\pi. \tag{5.44}$$

The speckle patterns will correlate when

$$\Delta\Theta = 2n\pi. \tag{5.45}$$

These regions with different correlation can be visualized as correlation fringes in different ways. In the *real time method* the specklegram of the surface before displacement is made and replaced in the original position after processing. Proper alignment of the positive ('live' image) and negative (processed specklegram) reduce the transmission to zero. Subsequent, non-uniform axial deformation of the object will give rise to fringes representing contours of equal phase difference, or constant displacement of the object [96]. The bright fringes will correspond to uncorrelated speckle patterns, and the dark fringes will correspond to the regions where the speckle patterns are correlated. Sometimes, this *real time* method is also cited as the *mask* technique [164].

Two *masks*, one for the speckle pattern before rough surface deformation, and the other after deformation, may be made by recording speckle patterns and processing the obtained specklegrams. The two plates, one containing the negative image of the speckle pattern, and the other the positive image of the another speckle pattern, are placed in register and adjusted to yield correlation fringes to an observer looking through the pair of plates [164].

The *double exposure method* is more frequently used to detect the positions of the correlation fringes. In this method both speckle patterns are recorded on a photographic plate before and after object deformation. The regions where the speckle pattern was formed as the addition of coherent speckles, have *different texture* as compared to regions where the speckle pattern was formed as an incoherent addition. In the regions with *speckle texture* produced by a coherent speckle field combination, the flux transmitted is higher than in the regions of incoherent combination of the speckle field [164]. The detection of the correlation fringes can be improved by spatial filtering of the double exposed specklegram; see the next chapter.

The *two-mask (or sandwich) method* has an advantage as compared to the *double exposure method* as it allows for the compensation of rigid-body movements.

Another attractive method for the observation of correlation fringes is based on the electronic subtraction of intensities before and after object deformation; see Section 5.2.3.

A comparison between *dual beam speckle interferometry* and *holographic interferometry* seems appropriate since both are interferometric methods measuring very small deformations. The stringent mechanical vibrational isolation required by hologram interferometry is also needed with dual beam speckle interferometry. However, the latter method is superior to holographic interferometry in the sense that fringes always localize on the object surface, and it allows the *in-plane* displacement to be measured independently. Furthermore, in the *speckle dual beam method* it is only necessary for the film to resolve the speckles, the size of which is proportional to the f-number of the imaging lens. Therefore by using a large value of the f-number, photographic film of very much lower resolution may be used. The speckle pattern can even be recorded and displayed by a high-resolution CCD-TV system, see [78].

One major drawback of the *speckle dual beam method* is the poor quality of speckle fringes. Also, the size of the object that can be studied is limited by the size of the collimating lens.

The dual beam speckle interferometer is applicable to transparent media studies in a manner quite similar to rough surface displacement measurements. To do this, a transparent medium can be inserted into the *object* beam before the rough surface, whereas the *reference* beam illuminates the rough surface directly. Both refraction on the density gradients and phase retardation due to refractive index path integrals may be recorded using the technique.

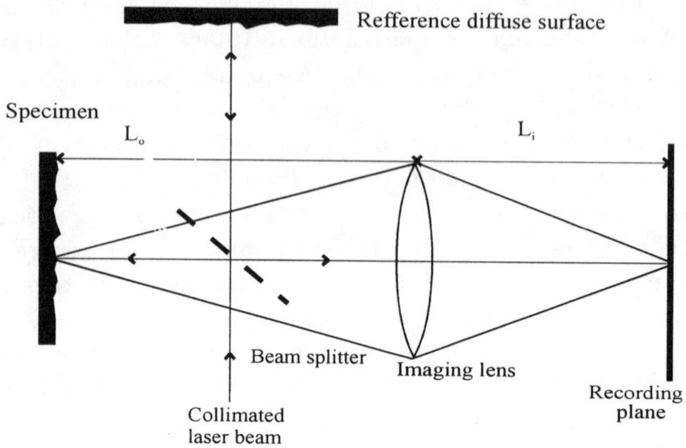

Fig. 5.10. Leendertz's dual beam speckle interferometer [161]

Another version of the dual-beam speckle interferometer for *out-of-plane* displacement measurements is shown in Fig. 5.10 [161]. The optical scheme is analogous to a Michelson interferometer in which both mirrors are replaced by scattering surfaces. One of the surfaces is the specimen while the other serves as a reference surface. The beam splitter permits both surfaces to be illuminated simultaneously by the incoming beam, and imaged simultaneously by the camera.

In this case, when the specimen is deformed, the relative phase change of the two speckle wavefronts is given by [164]

$$\begin{aligned} \Delta\Theta(x,y) &= \tfrac{2\pi}{\lambda}\cdot\Big[\tfrac{x\Delta_x}{L_0}\cdot\Big(1-\tfrac{x^2+y^2}{2L_0^2}\Big)+\tfrac{y\Delta_y}{L_0}\cdot\Big(1-\tfrac{x^2+y^2}{2L_0^2}\Big) \\ &\quad +2\Delta_z\cdot\Big(1-\tfrac{x^2+y^2}{4L_0^2}\Big)\Big] \end{aligned} \tag{5.46}$$

where L_0 is the distance from the specimen to the imaging lens.

For large *out-of-plane* displacements when $\Delta_z >> \Delta_x, \Delta_y$, the fringe equation for this interferometer simply reduces to

$$2\Delta_z\left(1-\frac{x^2+y^2}{4L_0}\right)=n\lambda. \tag{5.47}$$

This interferometer can be easily adopted for fluid mechanical measurements as well.

5.2.2 Speckle shearing interferometry

Generally, it is the strains that are of interest to stress analysis. The in-plane strains are related to the in-plane displacements by

$$\epsilon_x = \frac{\partial \delta_x}{\partial x}, \qquad \epsilon_y = \frac{\partial \Delta_y}{\partial y}, \qquad \gamma_{xy} = \frac{\partial \Delta_x}{\partial y} + \frac{\partial \Delta_y}{\partial x} \tag{5.48}$$

where ϵ_x and ϵ_y are normal strains and γ_{xy} is the shear strain. For measurements of these derivatives, a speckle shear interferometer is widely using. The principle of these methods is the same as for dual beam speckle interferometers, except that two interfering speckles now originate from the same object and are sheared laterally with respect to each other [164].

In Hung's shearing camera the lateral shear is introduced by means of a thin glass wedge placed in front of the imaging lens [89], see Fig. 5.11.

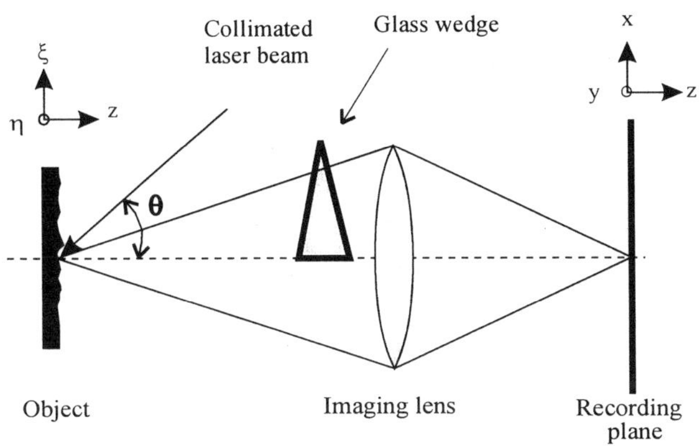

Fig. 5.11. Hung's speckle shearing interferometer [89]

The camera, which has a lens divided into two halves (one covered by a thin glass wedge), images the object, which is illuminated by a collimated coherent beam inclined at the angle Θ to the viewing direction as shown.

Since rays passing through the glass wedge are deviated, the two images (one focused by each half of the lens) are laterally sheared with respect to each other. In other words, the shearing camera brings the rays scattered from one point on the object into interference with those from a neighbouring point $P(x + \delta x, y)$. The shearing δx produced by the wedge is related to the wedge angle ϕ by

$$\delta x = L_0(\mu - 1)\phi, \tag{5.49}$$

where L_0 is the distance from the object to the wedge and μ is the refractive index of the wedge's glass. When the object is deformed there is a relative optical phase change $\Delta\Theta$ given by [164]

$$\begin{aligned}\Delta\Theta &= \tfrac{2\pi\delta x}{\lambda}\cdot\Big[\Big(\sin\theta+\tfrac{x}{L_O}-\tfrac{x^2+y^2}{2L_0^2}\Big)\cdot\tfrac{\partial\Delta x}{\partial x}\\ &\quad+\tfrac{y}{L_O}\cdot\Big(1-\tfrac{x^2+y^2}{2L_0^2}\Big)\cdot\tfrac{\partial\Delta y}{\partial x}+\Big(1+\cos\theta-\tfrac{x^2+y^2}{2L_0^2}\Big)\cdot\tfrac{\partial\Delta z}{\partial x}\Big].\end{aligned} \tag{5.50}$$

If the shear δx is small, the equation for the fringe contour reduces to

$$\delta x\cdot\left[(1+\cos\theta)\frac{\partial\Delta_z}{\partial x}+\sin\theta\frac{\partial\Delta_x}{\partial x}\right]=n\lambda. \tag{5.51}$$

The shear δx is equivalent to the gauge length. Should the shear be in the y-direction, the derivatives in Eq. (5.51) are with respect to y; this can be achieved by rotating the camera lens through 90°. If the object is rotated through 90° about the z-axis, Δ_x in Eq. (5.51) is replaced by Δ_y, the displacement component in the y-direction.

5.2.3 Electronic speckle pattern interferometry (ESPI)

Using electronic speckle pattern recording, it is easy to subtract the correlation fringes digitally, and to evaluate displacement, strain, etc., directly. This technique was introduced by Butters and Leendertz [76] and Macovski *et al.* [77] and is widely used today throughout the world [4,78-80].

The basic principles of ESPI have been thoroughly discussed elsewhere [4,23,80,169]. We will demonstrate here the possibilities of ESPI very briefly.

For displacement measurements the primary experimental data in speckle interferometry is the intensity distributions

$$I(x,y)=I_O(x,y)+I_T(x,y)+2\sqrt{I_O\cdot I_T}\cos(\Theta_O-\Theta_T). \tag{5.52}$$

The fringes analysis problem is to extract the optical phase difference, $\Theta=\Theta_O-\Theta_T$ and to determine the displacement.

The simplest way to do this is to measure the intensity distributions for different values of Θ_T using phase stepping [169,170]; thus acquiring for frames I_1, I_2, I_3, I_4 with $\Delta\Theta_T=\pi/2$, i.e. for $\Theta_{T1}=\pi/2, \Theta_{T2}=\pi, \Theta_{T3}=3\pi/4, \Theta_{T4}=2\pi$, the phase difference can be extracted by using the simple relation

$$\Theta_O(x,y)=\arctan\left[\frac{I_3-I_1}{I_4-I_2}\right]. \tag{5.53}$$

The phase contour is obtained here directly without correlation fringes analysis. Similar algorithms can be applied to different types of speckle interferometers [171].

6. Specklegram Treatment

Specklegram treatment involves mapping the displacement of a speckle pattern between exposures or during prolonged exposure. For photographic recording, this treatment can be performed by using point-by-point specklegram analysis or by optical processing of the whole specklegram. Using electronic speckle pattern recording, direct digital auto- or cross-correlation analysis of the recorded speckle patterns provides the sought displacements of speckles without traditional specklegram treatment. The same approach is used when a processed specklegram is digitized by electronic scanning or by using incoherent analysis of a microscopic portion of the specklegram. Some algorithms of such treatment will be described below as well as Young's fringes analysis in Section 6.1, whereas the optical processing of the whole specklegram and spatial filtering technique will be discussed in Sections 6.2 and 6.3 respectively.

6.1 Point-by-point specklegram analysis

In fluid mechanical applications, the speed with which a small area of the specklegram can be interrogated to obtain the local image displacement is a primary issue because the desired number of interrogation spots is always large. At present this specklegram treatment is performed with the help of automated image acquisition systems that allow one to input a Young's fringes pattern into a computer, to treat it, and to translate specklegram into the next interrogation point.

It is necessary to stress that during automated specklegram processing an enormous number of processed points can be obtained. This value is limited today only by the operation rate of the processing and by the amount of computer memory. Modern automated systems for specklegram processing are able to treat 10^4–10^5 specklegram points during reasonable processing time. It is understandable that the manual procedure of specklegram treatment, in spite of it simplicity, becomes impractical starting from a hundred processed points. It is interesting to note that a specklegram itself contains essentially more information than we could extract even with the help of the best automated treatment systems. Thus the standard size specklegram (9 by

12 cm^2) could be scanned with the step between neighbouring interrogation points equal to 10 μm to provide information in more than in 10^8 points.

A number of algorithms were developed both for the Young's fringes treatment, and for direct digital specklegram processing. The performance of the automated systems was thoroughly tested experimentally in many laboratories. The starting point of any automated system is the input of optical data into a computer. The image equitation is performed under illumination of the treated specklegram or by coherent laser light, or by normal white light, see Fig. 6.1. The former, *coherent methods*, require the specklegram to be examined point by point with a low-powered interrogating laser beam. This produces an interference fringes pattern whose analysis is used for the specklegram treatment. In the second method of specklegram analysis, a direct inspection is made of a portion of the image microscopically under *white light illumination* [172]. The specklegram microstructure is digitally recorded using a CCD camera attached to a framestore and processor, see Fig. 6.1.

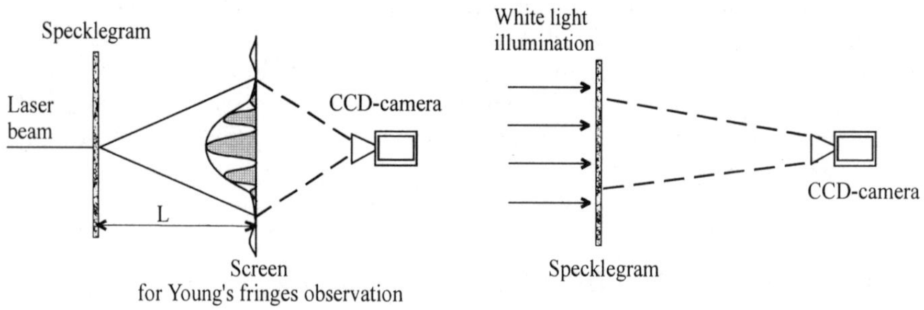

Fig. 6.1. Coherent (left) and incoherent (right) specklegram treatment

In contrast to the pointwise filtering scheme described below, a microscopic objective is used here for optical magnification of the analysed image. Thus the system of Heriot Watt University equipped with a Nikon microlens, has magnification[1] up to $M_2 = 20.6$ [172]. The portion of the film under examination was of the order of 1 mm^2. The image was digitized with the same resolution as in the Young's fringes analysis, namely a 1–256 grey level scene on a 512×512 pixel array using the CCD camera and framestore. Using a CCD matrix with dimensions 6.6×8.8 mm^2, the digital image corresponds to the flow dimensions 9.2×6.7 mm^2 at M_1 =0.05 and 3.1×2.0 mm^2 at M_1 =0.15 [172]. The last case corresponds to the flow resolution $R_x = 6$ μm/pixel

[1] The full magnification consists of two values, M_1 and M_2 where M_1 is the magnification from the object under study to the specklegram plane, whereas M_2 is the magnification from the specklegram plane to the CCD camera plane

and $R_y = 4\ \mu$m/pixel. After digitizing the image is available for numerical analysis.

Both methods, coherent and incoherent, have many advantages and disadvantages as compared to one another, and both techniques are effectively used for specklegram analysis.

In this section, after a short description of typical algorithms for Young's fringes analysis, a diagram of a fully automated system for *coherent* specklegram treatment will be derived.

6.1.1 Formation of Young's fringes

The most common approach to specklegram treatment is the measurement of the spacing and orientation of interferometric (Young's) fringes that are formed in diffraction halos under illumination of a small interrogation spot of the specklegram by a thin laser beam, see Fig. 6.2. Such specklegram treatment is cited as *point-by-point analysis* or *pointwise filtering.* Sometimes the technique is also called *coherent specklegram processing.*

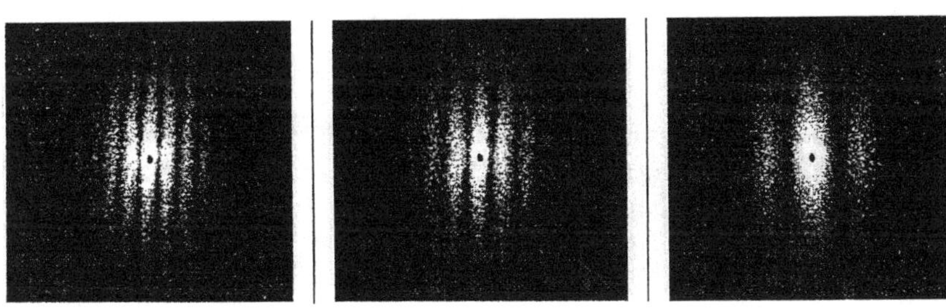

Fig. 6.2. Young's interferometric fringes for different speckle displacements

The intensity distribution in the diffraction zone behind the processed specklegram under illumination by a thin CW laser beam can be presented as [173,174]

$$I(x, y) = S(x, y) \cdot H(x, y) \cdot P(x, y) + N(x, y) + I_0(x, y) \tag{6.1}$$

where $P(x, y)$ describes the fringes system as created by the superposition of multiple speckle pattern exposures,

$$P(x, y) = 1 + V \cos[2\pi(f_x x + f_y y + c)], \tag{6.2}$$

with V being the fringes visibility, which may be position dependent,

$$V = \frac{I_{\max} - I_{\min}}{I_{\max} + I_{\min}}, \tag{6.3}$$

f_x, f_y being the fringes spatial frequencies and c being a constant. This constant may be position dependent also [175].

$H(x, y)$ reflects the signature of a single exposure, and is called *the diffraction halo*. $S(x, y)$ represents the strong multiplicative speckle noise component due to the coherent nature of image formation, whereas $N(x, y)$ introduces other additive noise contributions. $I_0(x, y)$ represents the zero-order spot located at the central part of Young's fringes. This spot may be blocked off easily by a simple stop-filter, see [176].

6.1.2 Filtering and removing the diffraction halo

Two negative specific features of Young's fringes pattern in speckle photography make the evaluation procedure difficult [176]. First, Young' fringes are superimposed on a strong speckle noise produced by the interference of the light, which is scattered from the specklegram and from the rough surface of the screen. The second is diffraction halo effects.

Due to the influence of the diffraction halo, fringe maxima do not exactly occur at the maxima of the cosine-squared fringes pattern $P(x, y)$. Therefore, the proper evaluation of the halo effects on the fringes pattern analysis is needed.

To suppress the additional speckle noise, both experimental and computational efforts were made. Erbeck [176] used a rotational ground glass as the screen. Thereby the interference conditions for forming the additional speckles vary with high speed, and as a result the fluctuating small-scale noise disappears. This noise is described by the S component in Eq. (6.1).

The noise component N in Eq. (6.1) can be reduced also through real time frame averaging in the image processor board. This method reduces, e.g., the noise added by the recording CCD camera and quantization noise from the analog-digital conversion [173].

Computational smoothing of the recorded images is often used to enhance the further processing as well as. Taking into account the radial symmetry of the Young's fringes pattern, Rösgen *et al.* [173] used the radial weighting functions (the windows), which are multiplied on to the raw experimental data. This has the effect of creating a circular image domain and makes the algorithm's performance independent of the fringes orientation. For sufficiently high fringes quality they used a Gaussian window of the form

$$W_{\mathrm{G}}(r) = \exp[-(r/r_0)^2], \tag{6.4}$$

whereas for low-quality fringes, they used a modified Hanning window

$$W_{\mathrm{H}}(r) = \begin{cases} 1, & r < r_0, \\ 0.5\left[1 + \cos\left(\pi\frac{r - r_0}{r_1 - r_0}\right)\right], & r_0 < r < r_1, \\ 0, & r_1 < r. \end{cases} \tag{6.5}$$

They have found that the optimal choice for r_0 and r_1 should be determined using the experimental set-up. For an image size of $N \times N$ pixels they used $r_0 = 3N/8$ and $r_1 = N/2$.

The first attempt to remove the diffraction halo prior to fringes analysis was made by Kaufmann [177]. It was found that considerable errors may arise if the diffraction halo in not removed before the fringes analysis. Later, the diffraction halo effects were extensively studied by many authors [178-189].

As the diffraction halo is the Fourier transform of a single-exposure image, the most practical way of removing the diffraction halo is experimental measuring of this distribution for a single exposure specklegram and dividing of Eq. (6.1) by the obtained $H(x, y)$. For the noise-free case and neglecting the constant c in Eq. (6.2), Eq. (6.1) can be re-written in the form:

$$\frac{I(x,y)}{H(x,y)} = 1 + V \cos[2\pi(f_x x + f_y y)]. \tag{6.6}$$

This operation could be performed both digitally, after the signal is digitized, and experimentally prior to that. Erbeck [176] performed this operation by using a special compensation plate with decreasing transmission toward its centre placed in the light path behind the specklegram. The best way to prepare such a compensation plate is to make a positive specklegram for a single exposure of the speckle pattern.

For digital processing we need to know the light distribution in the diffraction halo. As a first approximation, the diffraction halo, also referred to as *a pedestal*, was modelled as a centred Gaussian-like curve [190-192]

$$H(x,y) = \exp[-\pi(x^2 + y^2)/\sigma^2] \tag{6.7}$$

where σ is the particle image radius in PIV or speckle size in speckle photography.

For further approximation, it is attractive to know the exact theoretical dependence of the diffraction halo profile. It can be obtained by using the relations (3.39), (3.42) for the power spectral density derived by Goodman [22]. For the case of diffraction halo analysis the profiles were specially derived by Meynart [185].

In speckle photography, the diffraction halo is the *power spectral density* of a speckle pattern and is found to be equal to *the autocorrelation function* of the imaging aperture [22]. For the usual circular aperture the normalized diffraction halo is

$$H(\rho) = \frac{2}{\pi} \cdot \left[\arccos\left(\frac{\rho}{R}\right) - \frac{\rho}{R} \cdot \left(1 - \frac{\rho^2}{R^2}\right)^{\frac{1}{2}} \right] \tag{6.8}$$

where R is the halo radius, and $\rho^2 = x^2 + y^2$. For a square aperture with width L the diffraction halo is [22,185]

$$H(x,y) = \left(1 - \frac{xL_{\rm i}}{Lf}\right) \cdot \left(1 - \frac{yL_{\rm i}}{Lf}\right) \tag{6.9}$$

where $L_{\rm i}$ is the distance between the imaging lens and the image plane and f is the focal length of the lens. These diffraction halos are sometimes referred to as *coherent*, see [188]. In contrast, the diffraction halos for *incoherent* speckle photography are

$$H_{in}(x,y) = [H(x,y)]^2, \quad H_{in}(\rho) = [H(\rho)]^2. \tag{6.10}$$

The latter halos are the case for *white light speckle photography* [193], or for PIV records with isolated and non-interacting diffraction limited images of randomly distributed δ-scatters [185]. This corresponds to the case of *infinitesimal image elements* using the terminology of Cho [194].

For PIV records, the case of finite size circular image elements has been analysed in the literature as well [188,194]. A circular image element can be represented by means of the circle function

$$\mathrm{circ}(r) = \begin{cases} 1 & \text{for } r < 1 \\ 0 & \text{otherwise.} \end{cases} \tag{6.11}$$

The halo function in this case will be modified by the power spectrum of the circle function

$$H_{\rm fa}(\rho) = H(\rho) \cdot \frac{\mathcal{J}_1(2\pi\rho/R)}{\pi\rho/R} \tag{6.12}$$

with $\mathcal{J}_1$ being the Bessel function of the first kind and order 1. The finite size of the image elements decreases the fringes visibility with increasing image size [194].

For *incoherent* image addition, the latter modification function should be squared according to Eq. (6.10) [188].

6.1.3 Fourier transform techniques

The Fourier transform represents a very attractive method of fringes analysis which reduces the problem to a peak search in the power spectrum associated with the original fringes pattern 173,194-196].

The Fourier transform of the Young's fringes pattern $\mathcal{F}\{P(x,y)\}$ consists of a δ-function of unit amplitude at zero ($f_x = 0, f_y = 0$) and two δ-peaks of amplitude $V/2$ at frequencies $f(f_x; f_y)$ and $f(-f_x; -f_y)$.

In discrete form, the Fourier transform operator has the form

$$F_I(p,q) = \frac{1}{M}\frac{1}{N}\sum_{m=1}^{M}\sum_{n=1}^{N} I(m,n)\exp(-2\pi[mp + nq]). \tag{6.13}$$

Here $I(m,n)$ is the digitized fringes pattern modelled in Eq. (6.1) by the term $P(x,y)$. If the diffraction halo is not substracted prior to the Fourier transform, the obtained spectrum $F_I(p, q)$ is a convolution of the Fourier transforms of halo and fringes pattern:

$$F_I = \mathcal{F}\{H(x, y)\} \otimes \mathcal{F}\{P(x, y)\}. \tag{6.14}$$

The convolution operation simply places three replicas of $\mathcal{F}\{H(x, y)\}$ instead of the δ-function in the power spectrum. To take into account the diffraction halo effect, the replicas can be extracted from the power spectrum, see [173].

Such calculations have *a good noise immunity* (see [197]) but they need about $2N^2 \log N$ multiplications and additions per transform.

The reliability and accuracy of this algorithm was tested both experimentally and numerically [197,198]. It was shown that even for a relatively low image resolution of 64×64, the algorithm could deal with fringes visibility of under 10%, giving a standard deviation in displacement magnitude of 0.05 μm and an angle of $0.1°$.

The *power spectrum* of the Fourier transform can be constructed by using the relation

$$\mathcal{W}(f_x, f_y) = \sum_{m=0}^{M-1} \sum_{n=0}^{N-1} I(m, n)[c(n, f_x) - \mathrm{i}s(n, f_y)][c(m, f_x) - \mathrm{i}s(m, f_x)] \tag{6.15}$$

where

$$c(n, f_y) = \cos(2\pi n f_y / N)$$
$$s(n, f_y) = \sin(2\pi n f_y / N).$$

To avoid the 2D FT evaluation, Robinson [199] has proposed a simple and elegant way to determine the fringes inclination angle. Taking into account, that 'the main' bright fringe in the interferometric pattern is always centred, the sum of the intensity along the fringe centreline will be maximal among any other lines passing through the coordinate centre. So, by means of an iterative procedure of simple intensity summing, the angular fringes orientation with precision of an angular error less then $\pm 1°$ can be very rapidly evaluated prior the FT image analysis. Then, instead of the 2D FT, the 1D FT of the image can be performed in a direction orthogonal to the fringes orientation.

Such a simple reconstruction procedure is often used at present for specklegram evaluation and provides accurate and reliable data. The direct 2D FT is also often used for Young's fringes examination 195,198,200].

6.1.4 Walsh transform

To decrease the computational time, Huntley [195,201] investigated the possibilities of the Walsh transform for fringes pattern analysis. This transform is up to an order of magnitude faster than the Fourier transform due to the

Walsh functions taking only two amplitude values, +1 and −1. These Walsh functions are a complete orthogonal set and are given by [202][2]

$$\mathrm{wal}(u,t) = \prod_{r=0}^{p-1} (-1)^{u} p - 1 - r(t_r + t_{r+1}), \tag{6.16}$$

where u_r and t_r are the digits of the binary representation of u and t, respectively.

The 2D discrete Walsh transform of $f(m,n)$ is

$$W(p,q) = \frac{1}{N}\frac{1}{M}\sum_{m=0}^{M}\sum_{n=0}^{N} f(m,n)\mathrm{wal}(p,m)\mathrm{wal}(q,n). \tag{6.17}$$

The Walsh power spectrum can be constructed by using an analogy with the FT (see Section 6.1.3). Huntley [195] derived relations for this power spectrum and has shown that in spite of simpler calculations in the case of the Walsh transform its power spectrum has the same form as the Fourier transform power spectrum. This allows one to determine the fringes spacing and inclination.

6.1.5 Autocorrelation technique

The two-dimensional normalized *autocorrelation function* of the light intensity distribution in the Young's fringes pattern $R_I(p,q)$ is

$$R_I(p,q) = \frac{N \cdot M}{(N-p)\cdot(M-q)} \left[\frac{\sum\limits_{n=1}^{N-p}\sum\limits_{m=1}^{M-q} I(n,m)\cdot I(n+p,m+q)}{\sum\limits_{n=1}^{N}\sum\limits_{m=1}^{M} I^2(n,m)} \right]. \tag{6.18}$$

This *correlation function* (CF) of the Young's pattern consists of *the correlation fringes* which have the same pixel spacing as the original pattern. Therefore, the wavelength of the CF is a measure of the fringes spacing of the Young's pattern.

The wavelength can be determined by measuring the spacing of two crests of the 1D CF [176]. The 1D CF can be determined along different directions

$$R_{I1D}(p) = \frac{N}{(N-p)} \left[\frac{\sum\limits_{n=1}^{N-p} I(n,m)\cdot I(n+p,m)}{\sum\limits_{n=1}^{N} I^2(n,m)} \right]. \tag{6.19}$$

It should be noted that the same methods described above for *coherent* point-by-point specklegram analysis may be used for the *incoherent* analysis

[2] Additional information about the Walsh transform can be found in [203,204].

as well. Thus, the direct autocorrelation specklegram analysis will produce the *correlation peaks* which are the same as peaks in the frequency domain when the Fourier analysis of the Young's fringes is performed. Alternatively, the direct Fourier transform of a small 'interrogation' region of a specklegram produces *correlation fringes*, the same as Young's fringes in *coherent* specklegram probing. The latter is a simple consequence of the fact that the complex amplitude in Young's fringes pattern is the Fourier transform of the complex amplitude behind the interrogation spot on the specklegram, whereas the former is a consequence of the *autocorrelation theorem*, which connects the Fourier transform of the correlation function with the power (Wiener) spectrum distribution [25]

$$\mathcal{F}\{R_I\} = |\mathcal{F}\{I(x,y)\}|^2. \tag{6.20}$$

6.1.6 Crosscorrelation technique

The crosscorrelation technique will be described in Section 8.1.4.

6.1.7 Phase iteration method

Phase-shifting techniques are well developed for direct phase evaluation in electronic speckle pattern interferometry [169-171,205]. Recently, an analogue of this technique has been developed for Young's fringes pattern analysis [196,206]

Using the phase from the previous iteration, four phase-shifted fringes patterns are constructed by

$$I_i = 1 + \cos k[\phi_{n-1} + \frac{\pi}{2}(i-1)] \tag{6.21}$$

where n is the iteration index and k is a multiplication factor. For $k = 1$, for example, the constructed fringes pattern is the same as that of the original Young's pattern. If $k = 2$, the fringe number is doubled. After smoothing of these patterns by convolution with, e.g., a disk shaped window

$$I_i^{'} = I_i \otimes \text{circ}(r), \tag{6.22}$$

and the improved phase is calculated as

$$\phi_n = \arctan \frac{I_4^{'} - I_2^{'}}{I_1^{'} - I_3^{'}}. \tag{6.23}$$

As founded by Gu and Chen [196], in the case of Young's fringes pattern analysis, one or two iterations yield a phase which is a fairly good linear coordinate function with only local fluctuations. A least-squares fit is used in the next step to suppress the local fluctuations. A set of linear algebraic equations

$$a\sum x_i^2 + b\sum x_i y_i = \sum x_i \phi_i$$
$$a\sum x_i y_i + b\sum y_i^2 = \sum y_i \phi_i \qquad (6.24)$$

is used to find the coefficients a and b and to best fit a phase at each point of the area of interest:

$$\phi_i = ax_i + by_i. \qquad (6.25)$$

Using these coefficients, the fringes spacing can be evaluated by

$$d_{\mathrm{f}} = \frac{2\pi}{\sqrt{a^2 + b^2}} \qquad (6.26)$$

and the speckle displacements are defined by the formulas

$$|\Delta| = \sqrt{\Delta_x^2 + \Delta_y^2} = \frac{\lambda L}{M d_{\mathrm{f}}} \qquad (6.27)$$

with M being the magnification of the recording system and L being the distance from the specklegram to the screen for the observation of the Young's fringe.

6.1.8 Directional ambiguity

Due to identical conditions in recording each of the two exposures on a specklegram, it is no possible to identify the order of speckle pattern exposures. As a result, measuring the separation of the speckle pairs on the specklegram is not sufficient to determine the polarity of the speckle shift between exposures. The simplest way to resolve this directional ambiguity is to make an additional, well-controlled speckle shift between these two exposures. In *particle image velocimetry* the time interval between two exposures is very short and special efforts are needed for resolving this directional ambiguity; see Section 8.1.3.

6.1.9 Automated system for specklegram analysis

The problem of fringes analysis is common to many interferometric techniques. As was already mentioned, in fluid mechanical applications, a whole flowfield could be reconstructed and a large number of measurements is required. This makes manual techniques impractical. Automated fringes analysis is a well-developed direction in fluid mechanical measurements [74,207-209].

Application of this analysis for specklegram treatment is, fortunately, facilitated by a simple interferometric pattern in the case of speckle photography. Since the first publications in the field of speckle photography it was clear that Young's interferometric fringes, which are produced by probing the specklegram with a narrow laser beam in the far-field diffraction region, can be processed automatically. The far-field diffraction pattern consists of a speckled diffraction halo, modulated by cosine-squared interferometric fringes, Young's fringes. These fringes have a universal form not depending on the processes being studied. The fringes pattern has in fact a rather simple form: equidistant straight lines, the distribution of light in which is theoretically well defined. The result of a Young's fringes treatment gives us two values, namely, the fringes spacing (the distance between neighbouring fringes), d_{fr}, and the orientation of these fringes, which it is possible to describe by the angle of fringes inclination with respect to, for instance, the horizontal axis, β_f.

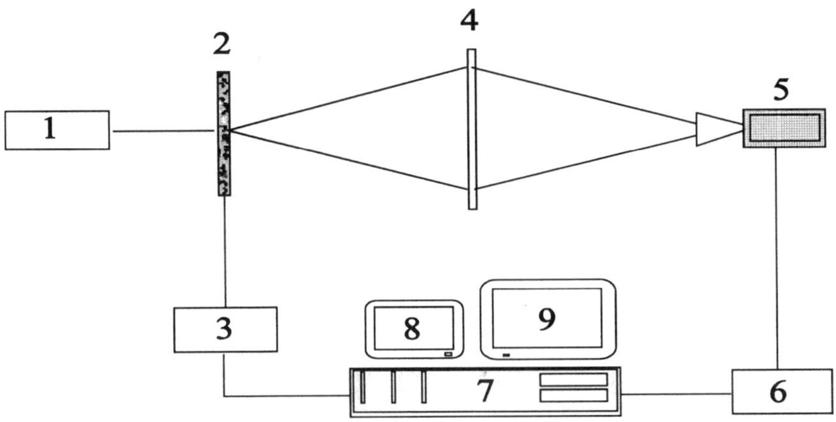

Fig. 6.3. Automated system for specklegram read-out and processing: 1, CW probing laser; 2, processed specklegram; 3, microtranslational stage with computer-controlled step motor; 4, screen for viewing the Young's fringes; 5, CCD camera; 6, interface; 7, host computer; 8, buffer storage; 9, monitor

The main components of an automated system for Young's fringe analysis is shown in Fig. 6.3. In general, the electronic components are the same as described above in Section 4.4; see Fig. 4.3. The specific parts are the system of Young's fringes formation and the specklegram translation. A narrow collimated CW low-power laser beam probes a small area of the developed specklegram, which is mounted on a microtranslational step motor driven stage, perpendicular to the laser beam direction.

The diffraction halo is formed on the screen situated behind the treated specklegram. A small stop mounted on the glass side in front of the CCD array reduces scattered light from the relatively high-power undiffracted beam. The displaced speckle pairs act as *coherent* diffraction sources, and interferometric fringes are produced in the diffraction zone.

A CCD video camera captures the intensity distribution in the halo and sends the information to a PC which determines the Young's fringes spacing and orientation, then stores this information together with the treated point coordinates and instructs the microtranslational stage to move the specklegram to another location. The frame contains 256 Kbytes of RAM for storing a 512 × 512 pixel image when the individual pixels are 8-bits deep.

This analysis process is repeated until the whole flowfield has been determined. Usually, complete analysis involves several thousand such measurements and so considerable interest has been shown in automated systems.

The first generation of systems for Young's fringes analysis used a semi-automatic regime, with the alignment of the fringes with the image device being performed manually, see [210-112]. In all these systems an electric read-out device was used for data input. Thus Maddux *et al.* [210] scanned the fringes using a photodetector, Ineichen *et al.* [212] used a TV camera and Kaufmann *et al.* [211] projected the fringes on to a linear photodiode array.

In the second generation of automated data processing systems a fully automated regime was realized, see [140,176,195,199,213-215]. The first fully automated specklegram evaluation systems were developed by Robinson [199], Toyooka *et al.* [215], and Erbeck [176].

In such systems the Young's fringes pattern is acquired by a TV camera and its output is transferred to the host computer as an image of 512×512 pixels with 256 grey levels. It has been demonstrated that such systems become a practical tool for specklegram processing providing very good precision of the automated speckle shift measurement [197].

6.1.10 Parallel processing

By using fast computers, the Young's fringes pattern analysis time is becoming lower than the time for mechanical translation of the specklegram to the next interrogation point. Taking this into account, Hinsch and Arnold [188,216] and Huntley [200] have developed extensions of the hybrid optical-electronic correlator, in that the speckle photographs are probed by an array of beams to form a 2D array of Young's fringes patterns. To form an array of beams, Hinsch and Arnold have used an array of small lenslets, which were fabricated holographically by sequential recording of small holograms of a converging beam at the grid points of the array [175]. Huntley used a crossed phase grating to form such a beam array. The lens is separated from the grating by the distance f, equal to the focal length of the lens. This method of forming coherent parallel beams is widely used in laser techniques for spatial laser light modulators [66].

The diffracted beams passing through the lens emerge parallel to one another. The observation plane position here corresponds to the Fourier Transfer (FT) plane of the specklegram placed in the array of beams. As usual, the specklegram is mounted on a stepper-motor-driven X-Y translational stage controlled by computer, which permits the specklegram to be moved after each set of fringes patterns is analysed. The fringes pattern is digitized by means of a CCD camera focused on the screen (a ground glass with individual stops to block out the undiffracted light for each beam) and analysed as described above.

The separation of the probe beams on the specklegram is determined by the focal length f and the spatial frequency of the grating which in the experiments of Huntley [200] were 160 mm and 75 lines/mm respectively and give a mesh spacing of 10.5 mm horizontally and 7.4 mm vertically. The beam separation should be chosen so that the region covered by the array is comparable with the required field of view. The specklegram need then be moved by only the beam diameter between successive frames. The diameters of individual halos on the observation screen is determined by the distance L and the speckle size σ, and must be less than the halo separation.

Using a Sun SPARC station 2 computer with benchmark ratings of 28.5 MIPS (million instructions per second) and 4.2 MFLOPS (million floating-point instructions per second), a CCD camera with 512×512 pixels which was subdivided into 16 images with 128×128 pixels, an evaluation time of 160 ms per frame was obtained by Huntley [200]. The total time taken for analysis includes all 'overheads' such as moving the specklegram to the next interrogation position.

This operation rate allows one to obtain millions of speckle displacement vectors using a finite treatment time. This opens real perspectives to evaluate complex 3D flows with a very good spatial resolution, comparable with the possibilities for computer modelling of such flows using the most powerful modern computers.

6.2 Optical processing of specklegram

Optical image processing refers to manipulation of images by optical means. In this way, an image can be decomposed into a spatial frequency or Fourier components. These components may be altered by filtering techniques. Thus *low-pass filtering* is achieved by placing an opaque disk with a centred hole in the back focal plane of the Fourier transforming lens, and *high-pass filtering* involves positioning an opaque disk on an axis at the location. *Band-pass filtering* is achieved by using transparent rings [30].

The attractiveness of optical image processing stems from the speed with which the signals are processed. A number of all-optical and hybrid optical-digital processors has been developed for automated specklegram treatment. We will describe below some of these schemes and will start from an *all-digital* processor.

6.2.1 All-digital processor

Going back to the incoherent specklegram analysis scheme shown in Fig. 6.1, on the left, let us describe the *all-digital* procedure of the treatment. As

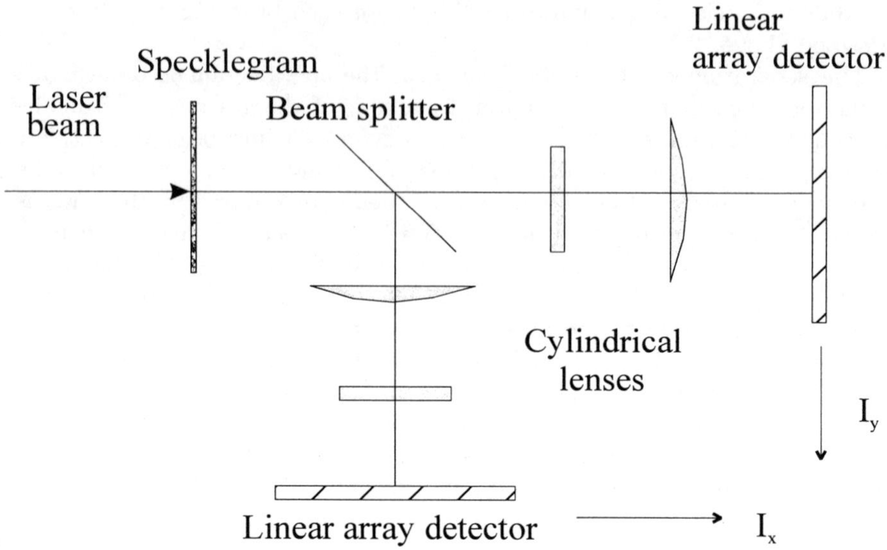

Fig. 6.4. Optical system for orthogonal compression of the interrogation spot image [218]

with many others, the procedure is based on the Wiener–Khintchine theorem, which may be expressed as follows

$$R(s_x, s_y) = \mathcal{F}_{\mathcal{A}}^{-1}\left\{|\mathcal{F}_{\mathcal{A}}\{T(x,y)\}|^2\right\} \tag{6.28}$$

where $T(x,y)$ is the transmissivity of the specklegram, $R(s_x, s_y)$ is its autocorrelation function, and the Fourier transform $\mathcal{F}_{\mathcal{A}}$ is taken of the product of the image transmissivity and the aperture function $\mathcal{A}$ [217]. As one can see, the digital procedure based on this concept involves a Fourier transform of the image, a squaring of this FT to obtain the power (Wiener) spectrum, and a second (inverse) FT of the power spectrum to obtain the autocorrelation function. This procedure needs about 10 s using a PC (486, 50MHz) including the detection of peaks of the autocorrelation function [217].

6.2.2 1-D optical compression

To reduce the computational time of the digital analysis of PIV records, Yao and Adrian [218] have proposed replacing the 2D light intensity distribution in the image plane $I(x,y)$ by two 1D distributions

$$I_x = \int I(x,y)\,\mathrm{d}y, \qquad I_y = \int I(x,y)\,\mathrm{d}x. \tag{6.29}$$

The integration was performed optically by using cylindrical lenses to image $I(x, y)$ on to a 1D array with image compression in the direction perpendicular to the array, as shown in Fig. 6.4. Each element of the array responds to the light flux integrated across its width, so that the scanned outputs of the arrays represent I_x and I_y. These signals are digitized and evaluated to obtain 1D correlation functions

$$R_x(s_1) = \int I_x(x) I_x(x + s_1)\,\mathrm{d}x, \quad R_y(s_2) = \int I_y(y) I_y(y + s_2)\,\mathrm{d}y. \tag{6.30}$$

These functions possess maxima at values of (s_1, s_2) corresponding to the mean particle displacement within the interrogation spot.

6.2.3 Hybrid optical–digital processor

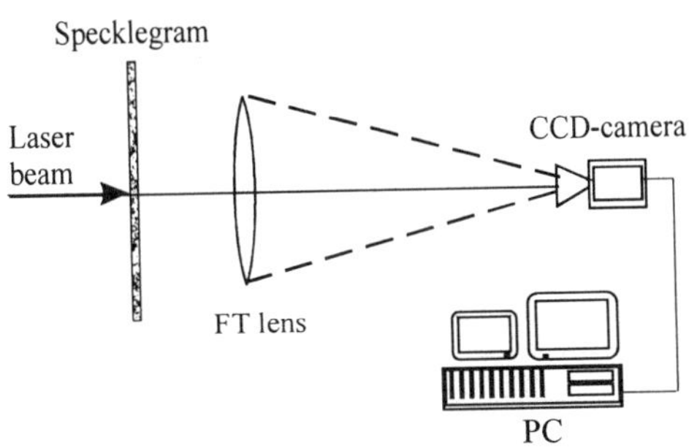

Fig. 6.5. Hybrid optical–digital processor

The *hybrid optical–digital processor* involves optical elements making an optical Fourier transform. The power spectrum of a small interrogation area of the specklegram is formed on the CCD array as shown in Fig. 6.5. This intensity distribution is the same as in the far diffraction zone and consists of Young's interferometric fringes; see the previous section. The intensity power spectrum is then digitized by a plug-in frame grabber. The digitised image is kept in memory and a digital fast (inverse) Fourier transform is performed, as described in Section 6.1.

6.2.4 Anamorphic optical processor

Collicott and Hesselink [219,220] have described an optical processor for the evaluation of speckle displacement optically along a selected line in the specklegram simultaneously without mechanical scanning. The schematic diagram of such optical system is shown in Fig. 6.6.

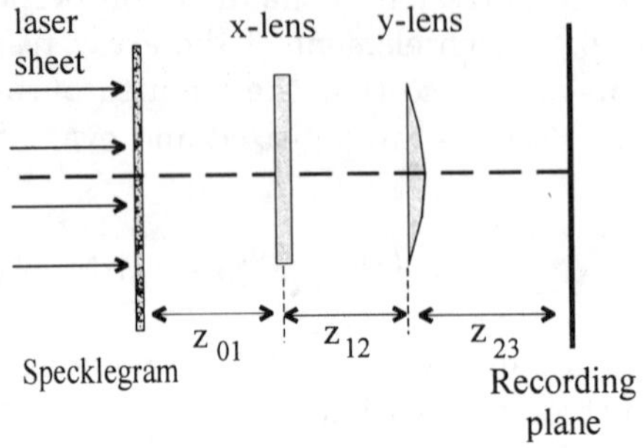

Fig. 6.6. Anamorphic optical processor [219]

A laser sheet, thin in the x-direction and extended in the y-direction, is used to probe a line of the specklegram of which the anamorphic optical system forms a one-dimensional Fourier transform in the x-direction and images of the speckle pattern in the y-direction for measuring the x-velocity component. This results in curved fringes, which have a local spacing inversely proportional to the x-component of the speckle displacement.

The focal length of the cylindrical lenses and distances should be chosen such that

$$f_x = Z_{01} = Z_{12} + Z_{23} \tag{6.31}$$

and

$$\frac{1}{f_y} = \frac{1}{Z_{01} + Z_{12}} + \frac{1}{Z_{23}}. \tag{6.32}$$

6.2.5 All-optical processors

Fully optical methods for analysing specklegram are promising alternatives to the digital treatment in the sense of speeding up the processing time. An *all-optical processor* may be implemented in several ways. The major design problem is the conversion of the power spectrum resulting from the detection of the first optical Fourier transform (OFT) into a coherent field, on which the second OFT is able to work [217]. The simplest solution of this problem is to record the Young's fringes pattern on a high-resolution film and to illuminate this film by coherent radiation after chemical ('wet') processing. Such way was used, e.g., by Arnold and Hinch in a parallel optical evaluation scheme [221]. To avoid 'wet' processing, a number of alternatives was proposed. Thus, methods using photorefractive materials [222-224], electrically addressed spatial light modulators (EASLMs) [225], and optically addressed spatial light modulators (OASLMs) [226,227] have been reported. The principle and high efficiency of optical processing has been demonstrated [217,228-230].

Both numerical and optical processors are based on algorithms using the Wiener–Khintchine theorem cited above. In the EASLMs based systems,

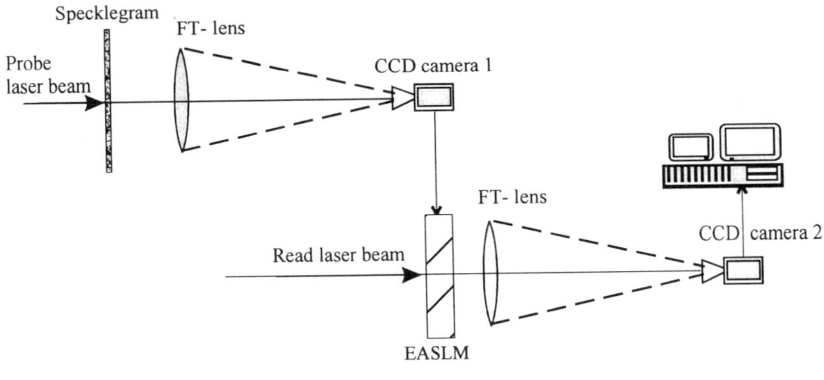

Fig. 6.7. All-optical processor controlled by video signal

Young's fringes patterns are formed when the specklegram is interrogated by *probe laser beam* and are captured on the first CCD camera. The video signal from this camera is transferred to the EASLM, which modulates the *read laser beam.* This field, which represents the power spectrum of the image, is Fourier transformed by the second FT lens and the result is picked up by the second CCD camera for subsequent digital processing; see Fig. 6.7. These systems have the same temporal and spatial resolution as CCD cameras, e.g. frequencies up to 30 Hz, and matrix size up to (1317×1035) [228].

A similar all-optical processor is based on a photorefractive crystal [222-224]. The crystal is illuminated by the Fourier transformed laser light passed through a small interrogation region of the specklegram. This field is stored as a hologram in the crystal. The *read beam* probes the crystal and the diffracted beam is Fourier transformed by the second FT lens. This results in the square of the autocorrelation function of the specklegram image [217]. This resulting beam is picked up by the CCD camera for subsequent digital analysis, namely, for the determination of peaks of the autocorrelation function. The typical write/read/erase times of such systems with a low-power Ar-ion probe laser are of the order of a few ms.

An example of an optical system based on the OASLM and developed at the Department of Physics and Astronomy of the University of Edinburgh [227] to perform such a specklegram treatment is shown in Fig. 6.8. The probe laser beam is split of by a beam splitter (BS) and directed on to the specklegram. The power spectrum of a small interrogation area of the specklegram is formed on the OASLM by the first FT lens. The OASLM detects the spatial intensity distribution of the Fourier transformed interrogation area and modulates the polarization state of a plane coherent light wave reflected from the back surface being a twisted nematic liquid crystal cell [227]. Passing the reflected light through an analyser the polarization modulation is turned into an amplitude modulation. The read laser beam is

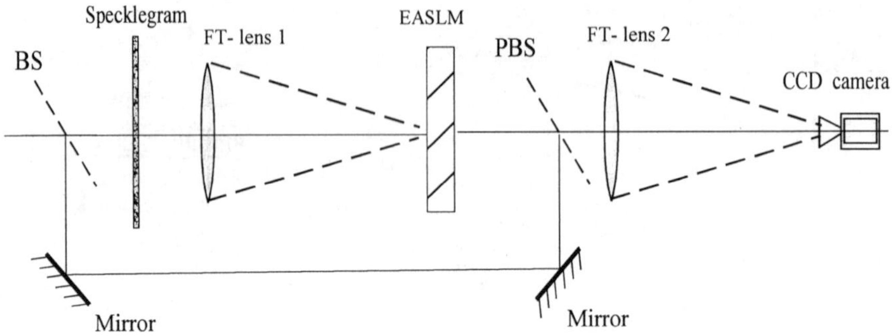

Fig. 6.8. All-optical processor of the University of Edinburgh, based on the OASLM [227]

formed with the help of the polarizing beam splitter (PBS) shown in Fig. 6.8. The reflected light is coherent and its amplitude immediately after the analyser (the PBS) will be modulated with the intensity distribution of the first Fourier transform. Thus, the intensity power spectrum is read coherently and Fourier transformed by the second FT lens to form the autocorrelation on the CCD camera. The CCD detects the squared autocorrelation function, which is then analysed digitally. The computer searches for the peaks of the autocorrelation function and based on their location determines the mean speckle shift within the selected interrogation area on the specklegram. By scanning the entire specklegram through the interrogation area the complete speckle shift $\Delta_s(x,y)$ can be measured.

Different types of commercial OASLMs may be used in such a processor. An analogue one is based on a nematic liquid crystal cell, which gives it an analogue and approximately linear transfer function within a limited range of input light intensity. A binary OALSM is based on a ferroelectric liquid crystal cell having a binary transfer function [230].

Similar to other all-optical processors, the speed of the processor is limited by the peak-search routines and the scanning systems. Mao *et al.* [229] describe a fast method of scanning of a specklegram using electronically programmable light modulators. A scan rate of 500 interrogation areas per second was achieved.

The OASLM based systems have been demonstrated with a resolution of 175 lines pairs per mm, frame rates up to 3 kHz, and large active areas of about 7 cm^2. The sensitivity of the OASLM was found to be about 170 nJ/cm^2 [226,228].

Recently, Dantec Measurement Technology announced a dedicated PIV processor, *FLOW MAP PROCESSOR 2000*, which is able to process six Mpixels per second per processor (of which four may be used in parallel) [217].

This would allow processing of a specklegram containing 10 000 interrogation areas with a resolution 256 × 256 pixels in 27 seconds.

6.3 Spatial filtering

Point-by-point specklegram analysis enables a large amount of quantitative data about speckle displacement to be obtained, which may be transferred into 2D distributions of velocity (PIV), density gradients (DESP), or temperature fluctuations (SPESP). Besides providing such data, the specklegram may also serve for flow visualization by optical or numerical generation of a direct visual pattern of the flow by *spatial filtering* (SF). So, speckle photography combines the advantages of whole field methods with the possibility of obtaining quantitative data in the points of interest. Even in the case when an automated system for specklegram treatment is not available, it is possible to obtain the whole field image if the flow studied by using a spatial filtering technique. This method of specklegram analysis has been used since the first publications on speckle photography; see [90,163]. The SF generates a fringe pattern corresponding to equal speckle displacement. The visualization of equi-velocity contours in speckle velocimetry has been reported by Celaya *et al.*, [231], Meynart [46], Suzuki *et al.*, [130], and Gärtner *et al.*, [232].

Köpf [14] has used the SF of the specklegram for laminar and Merzkirch and Wintrich [233] for turbulent flow visualization.

The *numerical spatial filtering* (NSF) technique was also used for flowfield construction by using the results of a quantitative specklegram treatment.

Detailed description of the SF for analysis of double and multiple exposed specklegrams is given by Palero *et al.* [234], whereas the enhancing of speckle contrast variation in single exposure speckle photography by *optical spatial filtering* (OSF) is described by Briers [107].

6.3.1 Optical filtering of double- and multiple-exposure specklegram

The schematic set-up for the OSF is shown in Fig. 6.9. The specklegram is illuminated with an expanded laser beam. The plane of the specklegram is imaged by a lens in the Fourier plane where a pattern of Young's interference fringes would be formed. An opaque screen is placed in this plane. The screen has a transparent hole or slit at an off-axis location. A small amount of the light passes through the hole, and a second lens forms another image of the specklegram in the plane of observation.

Since every point of the specklegram is illuminated by the expanded light beam, the screen is covered by a large number of overlapping systems of Young's fringes. Light may pass through the hole from such systems that have a maximum at the location of the hole. In this way the hole selects

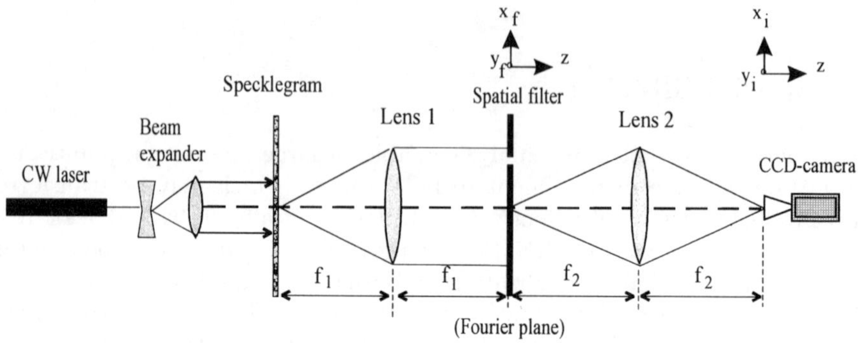

Fig. 6.9. Optical spatial filtering set-up

a particular spatial frequency and its harmonics which form the overlapping Young's fringes. Both in the case of *double exposure speckle photography* (DESP) and *multiple-exposure speckle photography* (MESP), the pattern can be a system of contour lines each of which represents a curve of constant speckle displacement.

Sometimes, a direct observation of the spatial filtering pattern is possible by looking at a specklegram under white light illumination [233]. In this case the pupil of the eye plays the role of the hole, and the pattern observed depends on the position of the eye with respect to the optical axis. Such a filtering process can also be performed by computation, once the specklegram has been evaluated via a quantitative point-by-point analysis.

A quantitative theory of spatial specklegram filtering is given by Palero *et al.* [234]. The starting point of the theory is the complex amplitude of the probing wave on the Fourier plane (FP), after the filter

$$\mathbf{U}(\mathbf{r}_\mathrm{f}) = \frac{\exp[4\pi\mathrm{i}f_1/\lambda]}{\mathrm{i}\lambda f_1}\mathcal{F}\{t(\mathbf{r}_\mathrm{o})\}P(\mathbf{r}_\mathrm{f}) \tag{6.33}$$

where $\mathbf{r}_\mathrm{f}$ is the position vector on the FP, $t(\mathbf{r}_\mathrm{o})$ is the transmittance of the specklegram, $P(\mathbf{r}_\mathrm{f})$ is the transmittance of the filter and $\mathcal{F}\{t(\mathbf{r}_\mathrm{o})\}$ is the Fourier transform (FT) of $t(\mathbf{r}_\mathrm{o})$:

$$\mathcal{F}\{t(\mathbf{r}_\mathrm{o})\} = \int_{-\infty}^{\infty}\int_{-\infty}^{\infty} t(\mathbf{r}_\mathrm{o}) \exp\left[-\mathrm{i}\frac{2\pi}{\lambda f_1}\mathbf{r}_\mathrm{o}\cdot\mathbf{r}_\mathrm{f}\right] \mathrm{d}x_\mathrm{o}\mathrm{d}y_\mathrm{o}. \tag{6.34}$$

The amplitude in the image plane is given by

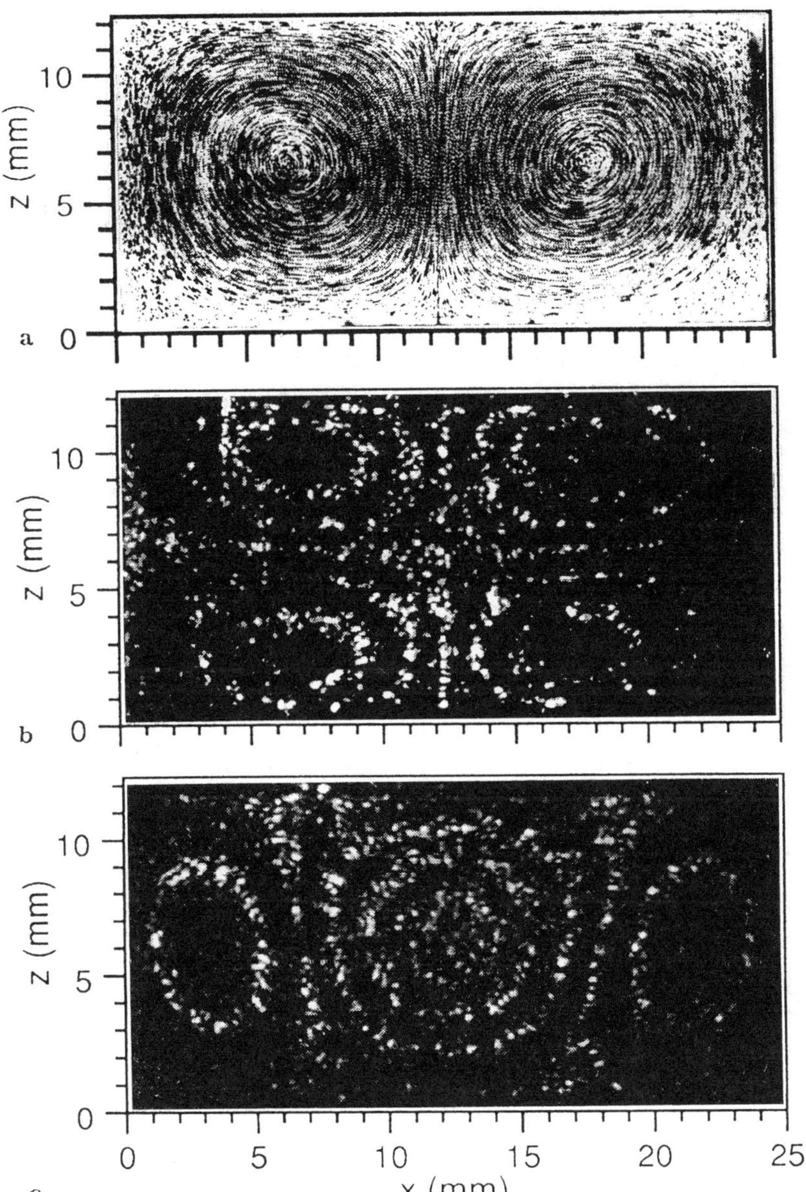

Fig. 6.10. Multiple exposure specklegram of Rayleigh–Bénard convective flow in a small rectangular cell (a) and the results of the optical spatial filtering (b,c). The increment between consecutive fringes is 25.0 μm/s for (b) and 34.0 μm/s for (c) (from Palero *et al.* [234])

$$\begin{aligned} \mathbf{U}(\mathbf{r_i}) &= \frac{\exp[4\pi i f_2/\lambda]}{\mathrm{i}\lambda f_1} \mathcal{F}\{\mathbf{U}(\mathbf{r_f})\} \\ &= \exp\left[\mathrm{i}\tfrac{4\pi}{\lambda}(f_1+f_2)\right] \tfrac{f_1}{f_2} \cdot t\left(-\tfrac{f_1}{f_2}\mathbf{r_f}\right) \otimes \mathcal{F}\{P(\mathbf{r_f})\} \end{aligned} \tag{6.35}$$

where $\mathbf{r_i}$ is the position vector on the image plane, and the FT of $P(\mathbf{r_f})$ is

$$\mathcal{F}\{P(\mathbf{r_f})\} = \int_{-\infty}^{\infty}\int_{-\infty}^{\infty} P(\mathbf{r_f}) \exp\left[-\mathrm{i}\frac{2\pi}{\lambda f_2}\mathbf{r_f}\cdot\mathbf{r_i}\right] \mathrm{d}x_f \mathrm{d}y_f. \tag{6.36}$$

Using these relations, Palero *et al.* [234] derived an equation for the intensity distribution in the image plane. If the spatial filter is a circular hole with radius R, whose centre is located at ρ, this equation is

$$\begin{aligned} I(\mathbf{r_i}) &= C \cdot \left| \int_{-\infty}^{\infty}\int_{-\infty}^{\infty} t\left[\tfrac{f_1}{f_2}(\mathbf{r}-\mathbf{r_i})\right] \exp\left[-\mathrm{i}\tfrac{2\pi}{\lambda f_2}\rho\cdot\mathbf{r}\right] \right. \\ &\quad \left. \times \frac{2J_1(2\pi Rr/\lambda f_2)}{2\pi Rr/\lambda f_2} \mathrm{d}x\mathrm{d}y \right|^2 \end{aligned} \tag{6.37}$$

where $C = (\pi R^2 f_1/f_2)^2$ is a constant.

For the multi-exposure specklegram (N-exposures) with equal speckle displacement between subsequent exposures, the specklegram transparency is

$$t = t(\mathbf{r_o}) \sum_{n=1}^{N} g[\mathbf{r_o} - (n-1)\mathbf{d}] \tag{6.38}$$

where $g(\mathbf{r_o})$ is the transmittance of the specklegram when only one exposure is taken, and $\mathbf{d} = \mathbf{d}(\mathbf{r_o})$ is the speckle displacement between two successive exposures, and Eq. (6.37) may be rewritten as

$$I(\mathbf{r_i}) = F(\mathbf{r_i}) I_h(\mathbf{r_i}) \tag{6.39}$$

where

$$F(\mathbf{r_i}) = \frac{\sin^2(N\pi\rho\cdot\mathbf{d}/\lambda f_1)}{\sin^2(\pi\rho\cdot\mathbf{d}/\lambda f_1)}$$

and

$$\begin{aligned} I_\mathrm{h}(\mathbf{r_i}) &= C \cdot \left| \int_{-\infty}^{\infty}\int_{-\infty}^{\infty} g\left[\tfrac{f_1}{f_2}(\mathbf{r}-\mathbf{r_i})\right] \exp\left[-\mathrm{i}\tfrac{2\pi}{\lambda f_2}\rho\cdot\mathbf{r}\right] \right. \\ &\quad \left. \times \frac{2J_1(2\pi Rr/\lambda f_2)}{2\pi Rr/\lambda f_2} \mathrm{d}x\mathrm{d}y \right|^2 . \end{aligned}$$

The *halo* term $I_\mathrm{h}(\mathbf{r_i})$ represents the image of the specklegram with a single exposure through the optical system. The *fringe* term $F(\mathbf{r_i})$ represents a fringe pattern that takes its maximum intensity, N^2, on the points where

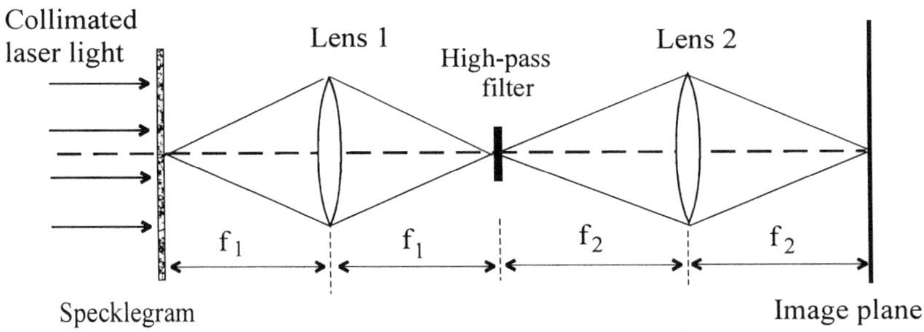

Fig. 6.11. High-pass optical spatial filtering

$$\frac{\rho \cdot \mathbf{d}(\mathbf{r})_{\mathrm{i}}}{\lambda f_1} = \frac{\rho d}{\lambda f_1} = m; \quad m = 0, \pm 1, \pm 2, \ldots \tag{6.40}$$

where d is the projection of the displacement vector on the direction defined by the filtering hole. Thus isodisplacement contours are drawn by the spatial filter, with the value for each fringe being

$$d = m \frac{\lambda f_1}{\rho} \tag{6.41}$$

and m being the order of the fringe.

The conditions for good fringe contrast in the spatial filtering technique were described in [234] and are very simple. They are

- R must be smaller than the least spacing of the Young's fringes in the Fourier plane, $R < \lambda f_1/d$;
- the fringe spacing, d_{f} in the filtered image must be larger than the speckle size, $d_{\mathrm{f}} > \lambda f_2/R$.

An example of the spatial filtering procedure is illustrated in Fig. 6.10.

6.3.2 Filtering of single exposure specklegram

For *single (prolonged) exposure speckle photography* (SPESP) the spatial filtering was applied by Briers to enhance speckle contrast variations in the PIV scheme [107]. The contrast of the speckle pattern for a prolonged exposure time is decreased in the region where flow is occurring due to blurring. The speckle pattern in an area of no flow will remain of high contrast. The blurring results from the fluctuations in the speckle produced by moving scatterers. In the line-of-sight scheme these fluctuations are produced when the light forming the speckle pattern is passed through turbulence [111-114,117].

The set-up for high-pass specklegram filtering is shown in Fig. 6.11. As in the OSF scheme, shown in Fig. 6.9, the specklegram is placed in a collimated laser beam. As described by Briers the higher the contrast of the speckles, the more light will be diffracted out of zero order. High-pass filtering is based on the suppression of zero-order light by a small circular stop-aperture.

7. Density (Temperature) Measurements in Laminar Flows

A light beam passing through the test field is disturbed due to the inhomogeneous distribution of the refractive index in the flowing fluid [12].

The speckle photography method is sensitive to the light deflection from its original direction and speckle interferometry, as well as classical interferometry, is sensitive to the phase shift of the disturbed light wave with respect to the reference wave. The speckle technique can be used both for flow visualization and for quantitative measurements of the density (temperature) fields in flows. The flow visualization approach in speckle photography is based either on the spatial filtering technique, using which one can obtain density or density gradient isolines, or on the numerical reconstruction of the flow field using quantitative data obtained about the flow. The advantage of the latter is based on the large amount of quantitative information about the flow studied, which can be read out and stored in the computer memory during automated specklegram processing.

As in the case of all 'classical' optical techniques, the important requirement for performing quantitative measurements is exact modelling of the ray tracing through the fluid density field, which will be discussed in the following. This modelling is based on the accumulated knowledge in the field of optical methods for flow visualization and measurements which have been summarized and reviewed in books and comprehensive articles, such as [8,9,12,235].

7.1 Rays tracing through refractive index field

The deflection of a light ray due to a density gradient can be estimated from geometrical optics. The main parameter is *the refractive index* determined as the ratio of the speed of light in vacuum to that in the medium. The refractive index is considered to vary as a function of the three spatial coordinates x, y, and z.

The behaviour of the refractive index is different for gases, liquids and plasma flows and will be briefly described below.

7.1.1 Refractive index in fluid flows

The refractive index of a fluid n and the fluid density ρ are connected by the Clausius–Mosotti relation [235]

$$\frac{n^2-1}{n^2+2} = \frac{\rho \mathcal{L} e^2}{3\pi m_e M} \cdot \sum_{(i)} \frac{f_\text{i}}{\nu_\text{i}^2 - \nu^2} \tag{7.1}$$

where $\mathcal{L}$ is the Loschmidt number, M is the molar weight of the fluid, e is the charge and m_e is the mass of the electron, f_i are the oscillator strengths with resonant frequencies ν_i, and ν is the radiation frequency.

Taking into account that the refractive index of most gases is very close to one (e.g., for air $n = 1.000292$), the Clausius–Mosotti relation can be simplified in the case of gases to the Gladstone–Dale relation

$$n-1 = \frac{\rho \mathcal{L} e^2}{3\pi m_\text{e} M} \cdot \sum_{(i)} \frac{f_\text{i}}{\nu_\text{i}^2 - \nu^2}. \tag{7.2}$$

The Gladstone–Dale relation is often written in the form

$$n - 1 = K \cdot \rho, \tag{7.3}$$

where the Gladstone–Dale constant K has the dimension of $1/\rho$ and depends on the characteristics of the gas as well as on the frequency of the radiation. The difference $(n-1)$ is also called *the refractivity*. Near to the resonant frequencies ν_i in the regime of *anomalous dispersion* the refractive index n becomes complex, and the imaginary part of n describes the absorption of the transmitted radiation in the gas. In this region the refractive index is many orders of magnitude greater than in the non-resonant frequency region [235]. Bershader [236] therefore proposed to improve the sensitivity of optical visualization methods by illumination of the gas flow with resonant radiation. Thus in studies with air at typical wavelengths in the visible region, $K = 0.23$ cm^3/g. The corresponding value for the resonant sodium vapour line at the peak of refractivity is $K = 6.28 \times 10^5$ cm^3/g [237]. This value is six orders of magnitude large than the non-resonant value. Later, a value of $K = 1.89 \times 10^5$ cm^3/g was determined at the half-power point of the sodium D_2 line with the help of a tuneable dye laser as a light source [238]. The results obtained verify that the refractivity has two maxima located at the half points of the absorption curve and that there is an enormous gradient of the refractivity near resonance lines. Away from the neighbourhood of the resonant frequencies, K is only weakly dispersive, as demonstrated for air in Table 7.1 [12].

The refractive index of a gas mixture composed of several constituents is given by

$$n - 1 = \sum_{(i)} K_\text{i} \cdot \rho_\text{i}, \tag{7.4}$$

Table 7.1. Gladstone–Dale constants for different gases

Gas	K, (cm^3/g)	Wavelength, (μm)	Temperature (K)
Air	0.2239	0.9125	288
Air	0.2250	0.7034	288
Air	0.2259	0.6074	288
Air	0.2274	0.5097	288
Air	0.2304	0.4079	288
Air	0.2330	0.3562	288
He	0.196	0.633	295
Ne	0.075	0.633	295
Ar	0.157	0.633	295
Kr	0.115	0.633	295
Xe	0.119	0.633	295
H_2	1.550	0.633	273
O_2	0.190	0.589	273
N_2	0.238	0.589	273
CO_2	0.229	0.589	273
NO	0.221	0.633	295
H_2O	0.310	0.633	273
CF_4	0.122	0.633	302
CH_4	0.617	0.633	295
SF_6	0.113	0.633	295

where the K_{i} are the Gladstone–Dale constants, and ρ_{i} are the partial densities of the individual components. If one defines the Gladstone–Dale constant K of the mixture by $n-1=K\rho$, with ρ being the density of the mixture, one has

$$K=\sum_{(i)} K_{\mathrm{i}}\left(\frac{\rho_{\mathrm{i}}}{\rho}\right), \tag{7.5}$$

where (ρ_{i}/ρ) designates the mass fraction of the components in the gas mixture [235]. The Gladstone–Dale constants of several gases are listed in Table 7.1.

Within a first-order approximation, the refractive index for a dissociated diatomic gas is

$$n-1=\rho\{K_{\mathrm{M}}(1-\alpha_{\mathrm{D}})+K_{\mathrm{A}}\alpha_{\mathrm{D}}\}, \tag{7.6}$$

where K_{M} and K_{A} are the Gladstone–Dale constants of the neutral or molecular gas and of the atomic gas, respectively. The ratio $K_{\mathrm{M}}/K_{\mathrm{A}}$ for oxygen is nearly one, whereas for nitrogen is very different from one [12].

The same procedure can be applied to determine the Gladstone–Dale constant of ionized gas. For a monatomic gas the refractive index is

$$n-1=\rho\{K_{\mathrm{A}}(1-\alpha_{\mathrm{I}})+K_{\mathrm{I}}\alpha_{\mathrm{I}}\}+N_e K_e^{'}, \tag{7.7}$$

where K_A and K_I are the Gladstone–Dale constants of the atomic gas and of the ion gas, respectively; N_e is the electron number density. The Gladstone–Dale constant of the electron gas, K_e' has dimension different from those of K_A and K_I. From plasma theory it follows for the refractive index of the electron gas [12]

$$n_e - 1 = -\frac{\lambda^2 N_e e^2}{2\pi m_e c^2} \tag{7.8}$$

with c being the velocity of light in vacuum, and λ, the wavelength of the light, which is transmitted through the gas mixture. If one measures λ in centimetres, one may also write

$$n_e - 1 = -4.46 \times 10^{-14} \lambda^2 N_e. \tag{7.9}$$

The refractivity for liquids is by several orders of magnitude larger than for gases. Therefore, even extremely small density differences in a liquid produce considerable changes in refractivity [239]. The behaviour of refractive index in water was carefully studied experimentally. At room temperature the index of refraction in clean water is $n_0 = 1.332156$ for T=20 C° at a wavelength of $\lambda = 0.6328\ \mu$m [12]. Dobbins and Peck [240] using a helium-neon laser found for the temperature range $20\,\mathrm{C}^\circ < T < 35\,\mathrm{C}^\circ$

$$n = n_0 + \Delta n \tag{7.10}$$

with Δn being

$$\begin{aligned} \Delta n \times 10^5 &= -8.376(T - 20\,\mathrm{C}^\circ) \\ &\quad -0.2644(T - 20\,\mathrm{C}^\circ)^2 + 0.00479(T - 20\,\mathrm{C}^\circ)^3. \end{aligned}$$

A density-stratified liquid is produced often by variation of the concentration of salt in water [241]. The refractive index for this mixture was measured as a function of density by a Michelson interferometer. For the density range 1.0 to 1.04 g/cm^3 the refractive index was found to be

$$n = 1.3322 + K(\rho - \rho_0) \tag{7.11}$$

with $K = 0.248\,\mathrm{cm}^3/\mathrm{g}$ and $\rho_0 = 0.9982\,\mathrm{g/cm}^3$. For more higher densities the Gladstone–Dale constant was found to be $K = 0.224\,\mathrm{cm}^3/\mathrm{g}$ [241].

7.1.2 Refraction of light in a density field

In terms of the geometrical optics approximation, the electromagnetic wave can be described by a real scalar function $\mathcal{E}(\mathbf{r})$, named the *eikonal* [242]

$$\mathbf{E}_0(\mathbf{r}) = \mathbf{e}(\mathbf{r}) \exp[ik\mathcal{E}(\mathbf{r})], \tag{7.12}$$

$$\mathbf{H}_0(\mathbf{r}) = \mathbf{h}(\mathbf{r}) \exp[ik\mathcal{E}(\mathbf{r})]. \tag{7.13}$$

The equation for the eikonal function can be obtained from the Maxwell equations (see Eq. 2.17) and it is the base equation for optical ray tracing:

$$(\frac{\partial \mathcal{E}}{\partial x})^2 + (\frac{\partial \mathcal{E}}{\partial y})^2 + (\frac{\partial \mathcal{E}}{\partial z})^2 = n^2(x, y, z) \tag{7.14}$$

The surfaces with $\mathcal{E}(\mathbf{r}) =$ Const are named geometrical wavefronts and can be described by a unit vector $\mathbf{s}$

$$\mathbf{s} = \frac{\text{grad }(\mathcal{E})}{|\text{grad}(\mathcal{E})|} \tag{7.15}$$

whose direction is coincident with the direction of electromagnetic wave propagation. Introducing the coordinate s along the optical ray so that $\mathrm{d}\mathbf{r}/\mathrm{d}s = \mathbf{s}$, we obtain the eikonal equation in the form

$$n\frac{\mathrm{d}\mathbf{r}}{\mathrm{d}s} = \text{grad}(\mathcal{E}) \tag{7.16}$$

where $\mathbf{r} = x\mathbf{i} + y\mathbf{j} + z\mathbf{k}$ is the position vector of a point on the ray.

This equation can be transformed to form which does not include the function $\mathcal{E}$

$$\frac{\mathrm{d}}{\mathrm{d}s}(n\frac{\mathrm{d}\mathbf{r}}{\mathrm{d}s}) = \text{grad}(n). \tag{7.17}$$

Weyl [243] derived an equivalent set of differential equations for ray tracing

$$\begin{aligned} \frac{\mathrm{d}^2 x}{\mathrm{d}z^2} &= \left\{1 + \left(\frac{\mathrm{d}x}{\mathrm{d}z}\right)^2 + \left(\frac{\mathrm{d}y}{\mathrm{d}z}\right)^2\right\} \cdot \left\{\frac{1}{n}\frac{\partial n}{\partial x} - \frac{\mathrm{d}x}{\mathrm{d}z}\frac{1}{n}\frac{\partial n}{\partial z}\right\} \\ \frac{\mathrm{d}^2 y}{\mathrm{d}z^2} &= \left\{1 + \left(\frac{\mathrm{d}x}{\mathrm{d}z}\right)^2 + \left(\frac{\mathrm{d}y}{\mathrm{d}z}\right)^2\right\} \cdot \left\{\frac{1}{n}\frac{\partial n}{\partial y} - \frac{\mathrm{d}y}{\mathrm{d}z}\frac{1}{n}\frac{\partial n}{\partial z}\right\}. \end{aligned} \tag{7.18}$$

For weak refraction, simplified equations may be derived. Thus in paraxial approximation and assuming, that the slopes of the ray, $\mathrm{d}x/\mathrm{d}z$ and $\mathrm{d}y/\mathrm{d}z$, are very small everywhere as compared with unity and the values $\frac{\partial n}{\partial x}$, $\frac{\partial n}{\partial y}$, $\frac{\partial n}{\partial z}$, are of the same order of magnitude, the simplified system of equations reads [12]

$$\begin{aligned} \frac{\mathrm{d}^2 x}{\mathrm{d}z^2} &= \frac{1}{n}\frac{\partial n}{\partial x} \\ \frac{\mathrm{d}^2 y}{\mathrm{d}z^2} &= \frac{1}{n}\frac{\partial n}{\partial y}. \end{aligned} \tag{7.19}$$

Due to the simplification introduced, the light ray leaves the test field at the same coordinates ξ, η (see Fig. 7.1) where it enters the field, but at an angle that is determined by the line-of-sight integral of the derivative of the refractive index distribution in the field.

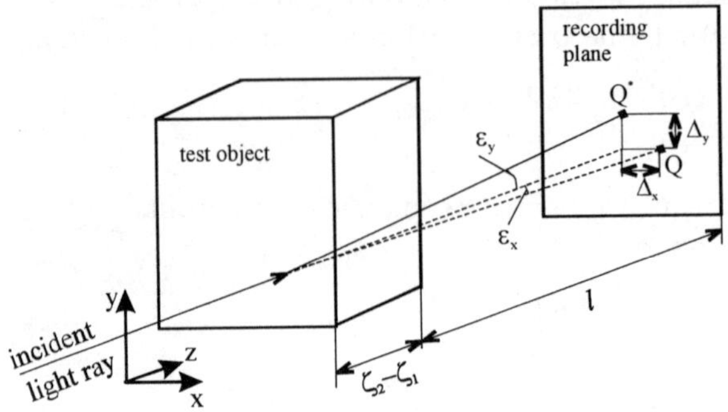

Fig. 7.1. Deflection of the light ray in an inhomogeneous test object

Denoting by (x, y, η_1) and (x, y, η_2) the coordinates where the ray enters and leaves the test volume correspondingly, one obtains

$$\Delta x = l \int_{\eta_1}^{\eta_2} \frac{1}{n}\frac{\partial n}{\partial x}\,\mathrm{d}z, \tag{7.20}$$

$$\Delta y = l \int_{\eta_1}^{\eta_2} \frac{1}{n}\frac{\partial n}{\partial y}\,\mathrm{d}z, \tag{7.21}$$

$$\tan\varepsilon_x = \int_{\eta_1}^{\eta_2} \frac{1}{n}\frac{\partial n}{\partial x}\,\mathrm{d}z, \tag{7.22}$$

$$\tan\varepsilon_y = \int_{\eta_1}^{\eta_2} \frac{1}{n}\frac{\partial n}{\partial y}\,\mathrm{d}z, \tag{7.23}$$

$$\Delta t = \frac{1}{c}\int_{\eta_1}^{\eta_2} \{n(x, y, z) - n_\infty\}\mathrm{d}z, \tag{7.24}$$

Here l is the distance of the recording plane from the exit plane of the test volume and Δt is the retardation of the disturbed ray with respect to the undisturbed one.

For a strong refraction, direct numerical integration of the ray equation may be performed. In [244] a more appropriate form of ray equation (7.17) was proposed for direct numerical integration. Defining a new variable t as follows:

$$t = \int \frac{\mathrm{d}s}{n}; \quad \mathrm{d}t = \frac{\mathrm{d}s}{n}, \tag{7.25}$$

Eq. (7.17) can be rewritten as

$$\frac{\mathrm{d}^2\mathbf{r}}{\mathrm{d}t^2} = n \cdot \mathrm{grad}(n) \tag{7.26}$$

or

$$\frac{\mathrm{d}^2\mathbf{r}}{\mathrm{d}t^2} = \frac{1}{2}\mathrm{grad}(n^2) \tag{7.27}$$

It is convenient also to use an optical ray vector $\mathbf{T}$ as proposed in [244]:

$$\mathbf{T} = \frac{\mathrm{d}\mathbf{r}}{\mathrm{d}t}. \tag{7.28}$$

Using these definitions Eq. (7.26) can be written as

$$\frac{\mathrm{d}^2\mathbf{r}}{\mathrm{d}t^2} = n \bigtriangledown n(\mathbf{r}), \tag{7.29}$$

or

$$\frac{\mathrm{d}^2\mathbf{r}}{\mathrm{d}t^2} = \frac{1}{2} \bigtriangledown n^2(\mathbf{r}). \tag{7.30}$$

The components of the vector $\mathbf{T}$ are optical direction cosines

$$\mathbf{T} \equiv \frac{\mathrm{d}\mathbf{r}}{\mathrm{d}t} \equiv n\frac{\mathrm{d}\mathbf{r}}{\mathrm{d}s} \equiv n\frac{\mathrm{d}x}{\mathrm{d}s}\mathbf{i}+n\frac{\mathrm{d}y}{\mathrm{d}s}\mathbf{j}+n\frac{\mathrm{d}z}{\mathrm{d}s}\mathbf{k} \equiv n\cos\alpha\mathbf{i}+n\cos\beta\mathbf{j}+n\cos\gamma\mathbf{k}, \tag{7.31}$$

where α, β, and γ are the angles the ray direction makes with the x, y, and z axes, respectively.

Let us consider the propagation of a ray that has initially angles α_0, β_0, and γ_0 with the coordinate axes.

Subsequently, the ray position and the slope are computed for a small increment in path length $\Delta t = 1/n\Delta s$ using an iterative procedure based on the Runge–Kutta method, described in [244]

$$R_{n+1} = R_n + \Delta t \cdot [T_n + \frac{1}{6}(A + 2B)] \tag{7.32}$$

$$T_{n+1} = T_n + \frac{1}{6}(A + 4B + C), \tag{7.33}$$

where

$$A = \Delta t \cdot D(R_n), \tag{7.34}$$

$$B = \Delta t \cdot D\left(R_n + \frac{\Delta t}{2}T_n + \frac{1}{8}\Delta t A\right), \tag{7.35}$$

$$C = \Delta t \cdot D\left(R_n + \Delta t T_n + \frac{1}{2}\Delta t B\right) \tag{7.36}$$

$$R = \begin{pmatrix} x \\ y \\ z \end{pmatrix}; \quad T = \begin{pmatrix} T_x \\ T_y \\ T_z \end{pmatrix} \equiv n \begin{pmatrix} \mathrm{d}x/\mathrm{d}s \\ \mathrm{d}y/\mathrm{d}s \\ \mathrm{d}z/\mathrm{d}s \end{pmatrix}, \tag{7.37}$$

and

$$D \equiv n \begin{pmatrix} \partial n/\partial x \\ \partial n/\partial y \\ \partial n/\partial z \end{pmatrix} \equiv \frac{1}{2} \begin{pmatrix} \partial(n^2)/\partial x \\ \partial(n^2)/\partial y \\ \partial(n^2)/\partial z \end{pmatrix}. \tag{7.38}$$

Using the notation introduced above the ray equation can be rewritten in the matrix form [285]

$$\frac{\mathrm{d}^2 R}{\mathrm{d}t^2} = D(R). \tag{7.39}$$

The optical path length (OPL) according to its definition is

$$W = \int n\,\mathrm{d}s = \int n^2\,\mathrm{d}t. \tag{7.40}$$

A convenient computational scheme for solving of this equation was developed recently [244,246]. It allows one to compute the ray positions and the slope for a small increment in the path length starting from the initial cross-section where the rays are parallel to the z-axis. This technique was used both for laminar and for turbulent flow studies [247,248] and some results will be given in chapter 9.

7.2 Density (temperature) measurements by speckle photography

The first demonstrations of the possibilities of speckle photography for density (temperature) measurements as applied to a transparent phase object were made in 1972, see [14-17].

Two different optical schemes shown in Figs. 5.5 and 5.6 were used in these experiments. In the experiments of Debrus *et al.* [16] and Mallick and Roblin [17] the ground glass acting as 'the speckle generator' is illuminated by a parallel beam of laser light and a photographic plate of high resolution is placed behind the ground glass at some small distance, d. As shown in chapter 5, the speckle shift due to the probing light deflection by angle ϵ is $\Delta = \epsilon d$. Double exposure speckle photography with one exposure being made without the object allows one to determine the field $\Delta(x, y)$, and therefore, the field $\epsilon(x, y)$. Using these data the temperature gradients were reconstructed in convective flows in the vicinity of a heated metallic rod [16] and in liquids using the simplified ray equations [143,249]. The spatial resolution of the measurements without an imaging lens is, of course, poor In spite of this the scheme was successfully applied to the study of free convection in a gap filled by liquid (silicone oil) and the results of temperature gradient measurements in the gap obtained using speckle photography were in good agreement with the data obtained by means of a thermocouple [143].

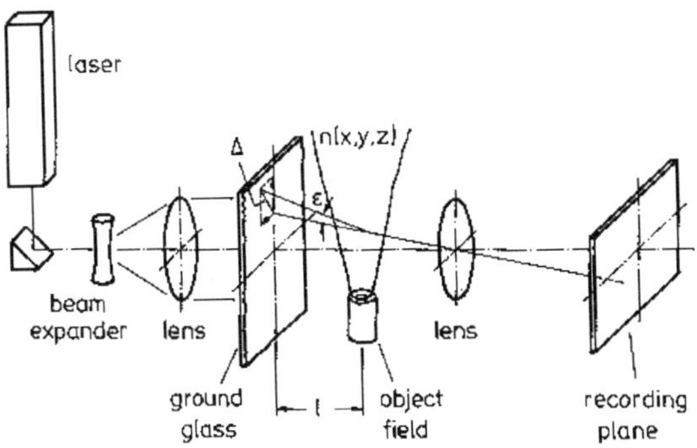

Fig. 7.2. Optical set-up for speckle photography of transparent objects

Figure 7.2 shows the optical scheme for obtaining specklegrams in which the transparent phase object under study is placed behind a ground glass. Such schemes were used in the experiments of U.Köpf, [14,15], and, later, in the publications [18,63,143,144,250,251], see Fig. 7.3. An expanded laser beam illuminates a plate of ground glass. The ground glass is imaged on to a photographic plate, and a first (reference) photographic exposure is taken in the absence of the test field. A second exposure is taken on the same photographic plate in the presence of the object under study. Again, due to the deflection of light in the test field, the speckle pattern recorded in the second exposure is displaced with respect to the reference pattern by the distance $\Delta(x, y) = \epsilon(x, y) \cdot l$

This scheme gives the same sensitivity of speckle photography on the deflection angle depending on the value l, what was described above for the scheme, in which the phase object is placed before the ground glass. A disadvantage of this scheme is the loss of applicability of paraxial approximation for the description of light refraction in the object.

The results show that double exposure speckle photography is a simple and is reliable means for measuring temperature gradients in transparent systems, and capable of measuring higher temperature gradients than is possible with holographic interferometry.

In [18] the temperature measurements in an open flame were made. It was shown that in the temperature range from $\sim$ 1000 K to $\sim$ 2400 K the data obtained by means of speckle photography are in excellent agreement with those obtained by a thermocouple traversed through the flame. Assuming the burner flame is axisymmetric the local values of temperature were reconstructed by applying an inverse Abel transform. The elimination of nu-

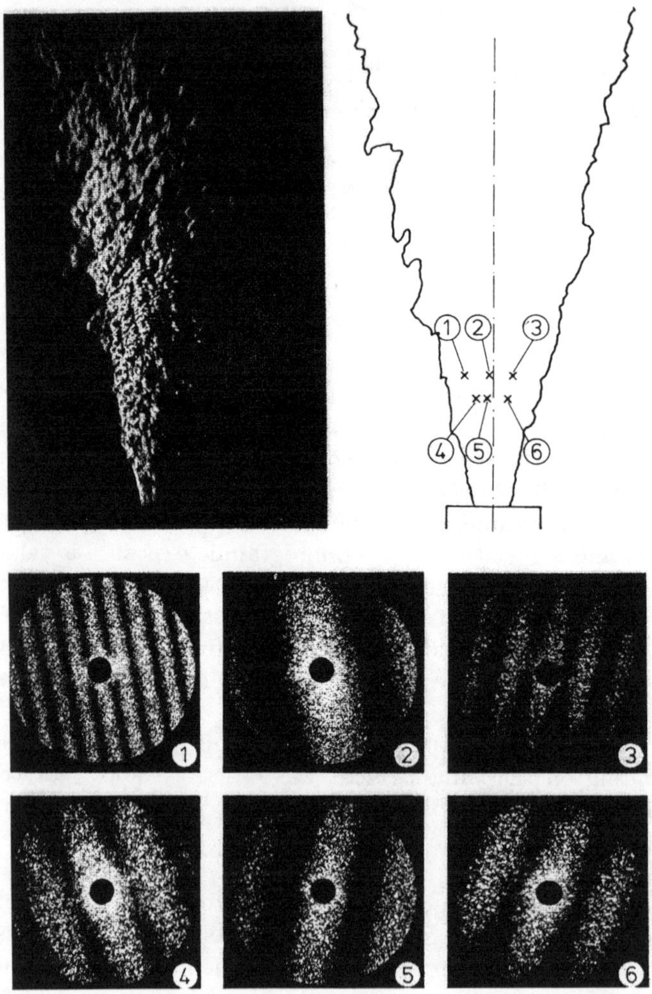

Fig. 7.3. Schlieren photograph of the turbulent jet. Six positions in the field of view are designated for which the respective systems of Young's fringes are shown (from Wernekinck, Merzkirch, and Fomin [144])

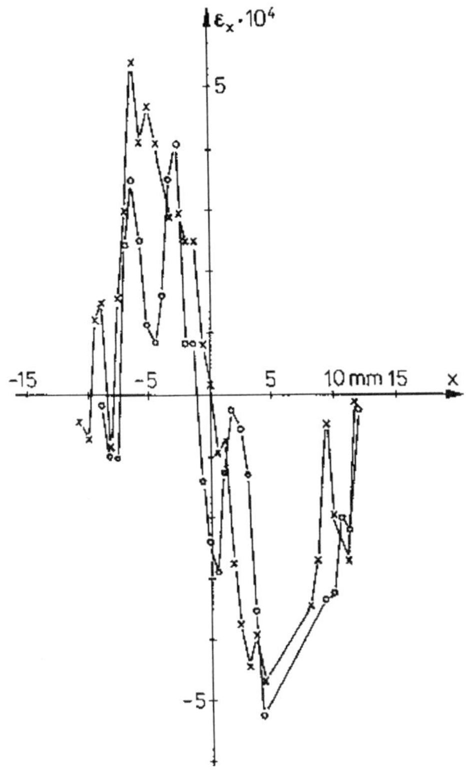

Fig. 7.4. x - component of the deflection angle, ϵ_x, measured along two horizontal lines across the field of view (from Wernekinck, Merzkirch, and Fomin [144])

merical differentiation of the experimental data to calculate the inverse Abel transform in the case of speckle photography was noted.

A turbulent jet was taken as an object in [144]. A pulsed ruby laser was used in these experiments to illuminate a ground glass in the scheme shown in Fig. 7.2. A schlieren photograph (Fig. 7.3) shows that the flow under study, the vertical helium jet, is completely turbulent. In the same figure, six positions in the field of view, i.e. six points (x_i, y_i), are marked for which the respective systems of Young's fringes are shown.

The different amounts of fringe spacing and direction indicate that the angle of light deflection is not regularly distributed. The distribution of the x-component of ϵ along two horizontal scanning lines in the field of view is seen in Fig. 7.4.

At the outer edge of the jet, the value of ϵ must approach zero. In the laminar case, ϵ would be negative on one side of the axis, positive on the other side, and zero on the axis itself. This regular pattern disappears for the turbulent jet; the deflection angle fluctuates with relatively large amplitudes.

The method adopted from speckle photography together with a short exposure time as realized by the ruby laser in a Q-switched mode allows for the measurement of the light deflection in a turbulent flow with fluctuating fluid density. The information about the density gradient is integrated along the optical path, and deriving the three-dimensional, instantaneous distribution of this gradient or the density itself appears to be a problem which can hardly be solved; see chapter 10. However, turbulent fields are usually described by certain statistical properties, rather than by quantities resolved in space and time. In chapter 9 a statistical analysis of the flowfield based on a fully automated specklegram evaluation system will be described.

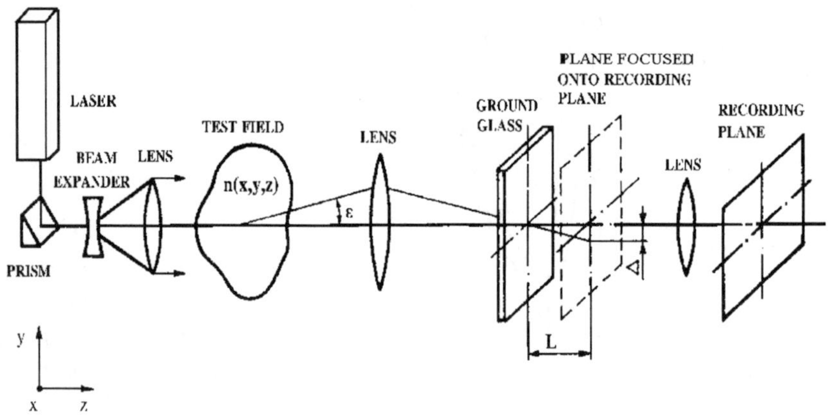

Fig. 7.5. Wernekinck and Merzkirch's optical set-up for speckle photography of an extended transparent object [252]

The optical system shown in Fig. 7.2 and used in these applications is limited to deflection angles that are not too large. Light deflection may be considerable in refractive index fields of large diameter or of wide extension in the direction of the traversing light beam. Then the photographic imaging through such an extended object is disturbed with respect to the reference exposure made in the absence of the object field. This results in a decorrelation of the two speckle patterns, the reference pattern and the object pattern [145].

Wernekinck and Merzkirch [252] introduced a novel optical arrangement which is widely used in speckle photography nowadays; see Fig. 7.5. The test object is placed in front (on the left side) of the ground glass, and it is imaged by means of a lens on to the plane of the ground glass. The photographic plate is focused on to a plane at distance L from the ground glass. As usual, two exposures are taken: a reference exposure in the absence of the test field, and a second (object) exposure with the test object in place, as shown in Fig. 7.5. Due to the imaging of the test object on the ground glass, a light ray, deflected by an angle ϵ in the test field, will arrive at the ground glass at the same point where the corresponding, undisturbed ray of the reference exposure has arrived [145]. The two rays form an angle ϵ/M_1 between each other, where M_1 is the magnification of the first lens. The relation between the speckle displacement at the object plane Δ and the light deflection angle is $\Delta = \epsilon L/M_1$, whereas the speckle shift recorded on the specklegram is $\frac{\epsilon L M_2}{M_1}$ with M_2 being the magnification of the second (imaging) lens. This speckle shift can be determined, e.g., with the point-by-point reconstruction technique described above (see chapter 6).

This optical system was applied by Wernekinck and Merzkirch [145] to the laminar free-convective flow of air along a vertical heated flat plate. Experimental results in the form of the non-dimensional temperature profile $\theta(\eta)$ are shown in Fig. 7.6.

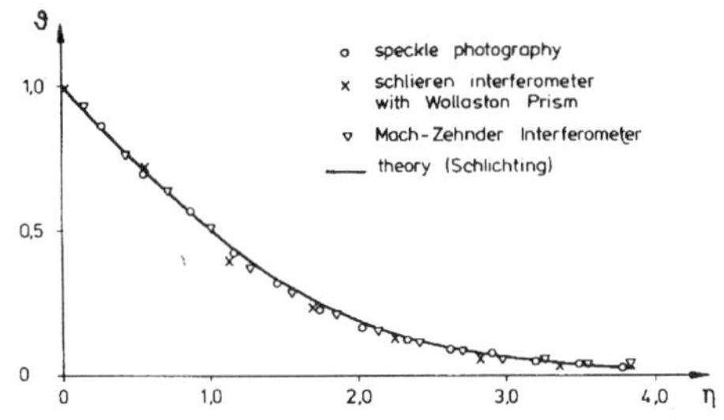

Fig. 7.6. Non-dimensional temperature in the laminar free-convection flow of air along a vertical, heated plate, plotted against the non-dimensional distance from the wall (from Wernekinck and Merzkirch [145])

The results are compared with the data obtained with a Mach–Zehnder interferometer and with a schlieren interferometer using a Wollaston prism. As it is possible to see, the agreement is excellent.

The further detailed comparison of the data obtained using speckle photography with those obtained by other optical methods was made by Wernekinck [63]. The comparison show that the precision of the data obtained with the help of speckle photography is very close to that of the other optical methods. The advantages of speckle photography over other optical methods noted by investigators at this period of time (before the widespread use of automated systems for specklegram treatment) were simplicity of the optical scheme, the possibility to obtain the vector values of interest, the capability of measurement of very high values of deflection angles, a relaxed requirement for mechanical stability of the optical components of a measurement scheme, *etc.* But the principal advantages of this technique over the others were realized later, when a wide variety of automated systems for specklegram processing was constructed and built in many laboratories over the world (see chapter 6). It should be noted that using automated specklegram processing quantitative data in hundreds of thousands points of the flow under study were obtained with a high precision of measurement. This 'density' of precise experimental information opens up new possibilities for the use of statistical analysis of these data for turbulent flows (see chapter 9).

7.3 Current trends in speckle photography of fluid flows

The high signal density that can be achieved in speckle photography makes this method particularly attractive for tomographic studies of three-dimensional flows and for the analysis of turbulent flows with density fluctuations. These directions in speckle photography are under development at Essen University (see, e.g. [19]) as well as at the Heat and Mass Transfer Institute (Minsk) (see, e.g. [20,253]) and will be discussed in the following chapters.

In addition to this, flow visualization by spatial specklegram filtering is being investigated at Essen University [233]. The technique was described in the previous chapter. As applied to flow with density variations, contour lines of constant displacement or constant light deflection angles can be observed under spatial filtering of the specklegram in the case of laminar flows, while the turbulent flow pictures are similar to schlieren records but with a lower degree of spatial resolution [19].

In series of publications from the Texas A& M University [147-150,254,255] natural convection flows were investigated by using speckle photography. Thus, convective heat transfer from an upward-facing isothermal plate has been studied by measuring the temperature gradients. It was shown that laser speckle photography can successfully determine local convective heat transfer coefficients in thermal flows.

An alternative way of flow visualization in speckle photography is to combine the set-up with an interferometer or with a schlieren system. Thus, Shu and Li [256,257] describe an arrangement that combines the speckle photography scheme with that of a schlieren interferometer, so that an 'on-line' visualization of the flow becomes possible. A pulsed ruby laser serves as the light source for speckle photography, whereas a CW He-Ne laser is used for observing the interferometric fringes in real time. An unstable flame of a Bunsen burners, as well as a laminar free-convective flow, were investigated using the combined set-up.

As far back as in 1967, R.I.Soloukhin used an optical interferometer of Michelson type, operating in a chronographic regime and equipped with a helium-neon laser as a light source, to measure the rate of oxygen recombination in shock waves [258]. In the framework of joint research project of the Heat and Mass Transfer Institute (HMTI), Minsk, with the University

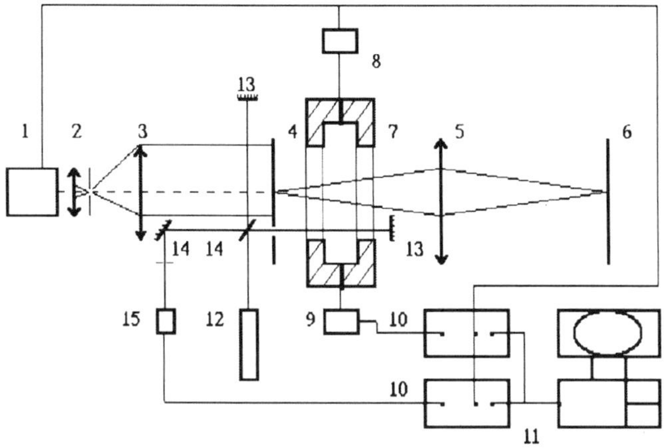

Fig. 7.7. Schematic diagram of installation for simultaneous digital Michelson interferometry and speckle photography. 1-pulsed laser; 2,3-collimator; 4-ground glass; 5-imaging lens; 6-photoplate; 7-combustion chamber; 8-spark plug; 9-pressure gauge; 10-oscilloscope; 11-computer; 12-He-Ne laser; 13,14-mirrors of the Michelson interferometer; 15-photomultiplier

of Poitiers, France, a novel combination of the Michelson interferometer with speckle photography (SP) measurements was made. This technique was applied to studying the knocking processes in the internal combustion engine, and was based on high-speed photography of interferograms with simultaneous photoelectric registration of the fringe shift at several fixed points [53,259].

The temperature evolution before the flame front, which is due to quasi-adiabatic compression of the fuel mixture, was identified by using the Michelson interferometer with digital recording. An independent method was employed for measuring the density drop in the combustion wavefront using a new modification of double-exposure speckle photography (DESP); see Fig. 7.7. Typical digital records of the Michelson interferograms at different flow positions are shown in Fig. 7.8.

A small optical wedge exists between the combustion chamber windows. Due to its being filled with gases of different densities, the ratio of the light deflection angles has the form

$$\frac{\varepsilon_1}{\varepsilon_2} = \frac{n_1 - n_0}{n_2 - n_0} \tag{7.41}$$

where ε_1 is the deflection angle in the region before the combustion wavefront, ε_2 is that behind the combustion wave front, n is the refractive index of

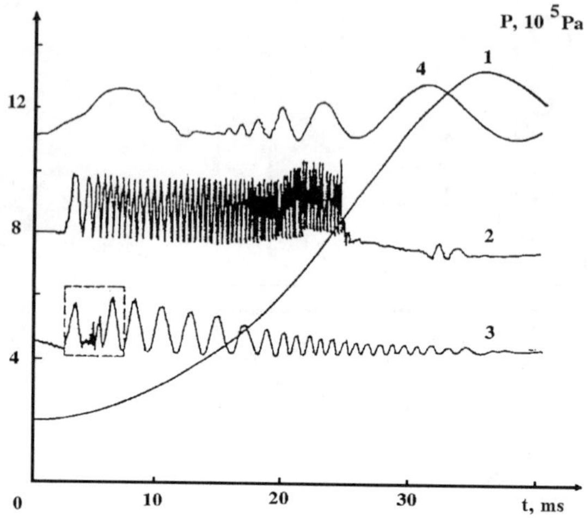

Fig. 7.8. The time dependence of pressure (1) and fringe interference, registered on the axis of the combustion chamber at a distance of 72.5 (2) and 2 (3) mm from the ignition spark.

the gases, and n_0 is the refractive index of the initial mixture. Using the Gladstone–Dale relation, we obtain from (7.41)

$$\frac{\varepsilon_1}{\varepsilon_2} = \frac{K_1 \cdot (\rho_1 - \rho_0)}{K_2 \rho_2 - K_1 \rho_0} \tag{7.42}$$

where ρ is the gas density. Relations (7.42) allows the reconstruction of the burnt mixture density from the known ratio of the light deflection angles in the optical wedge region and from the gas density before the combustion wavefront. The deflection angles were measured using double-exposure speckle photography. The holographic plate was exposed twice. During the first exposure, a stoichiometric propane-air mixture with the initial pressure P =1bar was located in the chamber. The second exposure was carried out within 15ms of the ignition instant. Variations in the densities of the burnt and unburned mixtures in the optical wedge region led to the deflections $\varepsilon_1, \varepsilon_2$ of the light beams, forming the speckle field, therefore, to its displacements $\triangle_1$ and $\triangle_2$. In this case

$$\frac{\triangle_1}{l_1} = \varepsilon_1, \quad \frac{\triangle_2}{l_1} = \varepsilon_2 \tag{7.43}$$

where l_1 is the distance between the ground glass surface and the combustion chamber end. The values of the $\triangle_1$ and $\triangle_2$ were measured according to the scheme of specklegram processing by measuring the spacing δ between Young's interference fringes:

$$\triangle_1 = \frac{\lambda L}{\delta_1}, \quad \triangle_2 = \frac{\lambda l}{\delta_2} \tag{7.44}$$

Here λ is the wavelength of the reconstructing laser beam, L is the distance between the specklegram and the screen for observing the interference fringes. It follows from (7.44) that

$$\frac{\varepsilon_1}{\varepsilon_2} = \frac{\delta_2}{\delta_1}. \tag{7.45}$$

Thus, the DESP with slightly inclined windows allows one to determine the density drop in the combustion or shock fronts. For the case of shock front diagnostics this technique was applied in [260].

To increase the sensitivity of speckle photography to a change in index of refraction, wavefront comparison (interferometric) systems have been tested [261,262]. Both photographic and electronic recordings may be used in speckle interferometry. The positive side of speckle interferometry as compared to holographic interferometry is the possibility to do the fringe visualization without an intermediate holographic print [4].

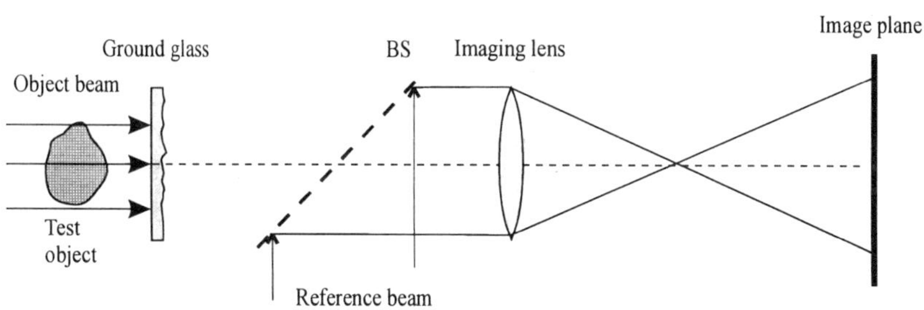

Fig. 7.9. Speckle interferometer for density measurements (from Verhoeven and Farrel [261]

The typical arrangement for speckle interferometry of a transparent phase object is essentially the same as for out-of-plane displacement measurements [23], but with the diffuse surface replaced by a ground glass diffuser, see Fig. 7.9. A camera lens is arranged to focus an image of the speckle generator surface on to the film. A second beam, a collimated reference beam, is combined with the scattered light at the beam splitter and is also imaged on to the film.

The first exposure, similarly to speckle photography, is taken in the absence of the object being studied. The surface of the diffuser is imaged on to a photographic plate along with the collimated reference beam generating a speckle pattern at the film plane. A second exposure is taken with the test object. The beam which passes through this object is distorted in phase, generating a speckle pattern which is similarly distorted.

In regions where the phase shift due to distortion in the object under investigation is equal to values of $2n\pi$ rad, the speckles of each exposure will be in phase and the speckle statistics for such regions will be correspond to the case of coherently combined speckle patterns (see Eq. 3.12):

$$p_1(I) = \frac{1}{< I >} \exp \frac{-I}{< I >}, \tag{7.46}$$

where $< I >$ is the averaged irradiance. In regions where the phase shift exhibits values of $(2n+1)\pi$ rad, the probability density distribution for the speckle brightness will correspond to the case of a sum of uncorrelated speckle patterns (see Eq. 3.24):

$$p_2(I) = \frac{4I}{< I >^2} \exp \frac{-2I}{< I >}, \tag{7.47}$$

This probability density function exhibits a most probable brightness, which is a considerably higher intensity than that of Eq. (7.46). As a result the combination of the speckle fields will give a pattern where regions corresponding to sums of correlated and uncorrelated speckles will be formed. Although both distributions (7.46) and (7.47), being normalized, give the same mean intensity, they can produce a different blackening effect on a photographic emulsion [90]. It was shown in [90] that the fringe visibility can reach values of more than 50 per cent for high contrast emulsions like Agfa-Gevert 10E70.

It should be noted that refraction effects in a speckle interferometer will decrease the visibility of the correlation fringes. The fringes can disappear when the lateral speckle shift between two exposures due to refraction in the object studied becomes comparable to the speckle size. A greater displacement than this can, however, be measured by the speckle photography technique.

In order to diminish refraction effects, an additional lens was introduced into the speckle interferometer scheme [262]. The lens is located between the test object and the ground glass so that it projects the light beams propagating through the density field. The test object was placed in these experiments in front of the ground glass. Electronic recording of the speckle pattern was used. 2D temperature profiles were calculated in a free convection flow. Song *et al.*, [263-265] have used a similar optical scheme with an additional lens and the test object placed behind the ground glass; see Fig. 7.10.

They also used electronic speckle pattern recording. A proper adjustment of the imaging lens allows one to diminish the speckle displacement effect and to improve the quality the interferometric fringes. The temperature fields in an acetylene and candle flame as well as in the natural convection zone near a heated vertical cylindrical tube were reconstructed. A speckle shearing interferometer with Wollaston prism and a phase shift interferometer were used by the same research team for the measurement of temperature in thermal flows [264,265].

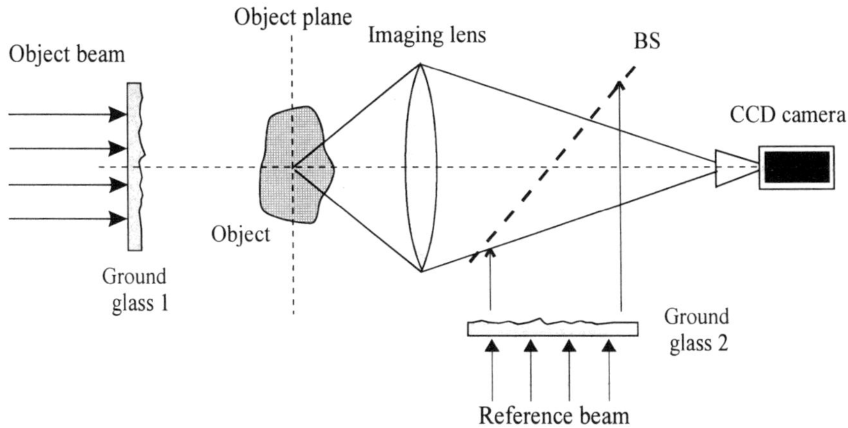

Fig. 7.10. Speckle interferometer for phase object study (from Song *et al.* [263])

Shakher *et al.*, [266-269] performed temperature profile measurements by speckle photography, speckle shearing interferometry and Talbot interferometry. They used a double aperture speckle shearing interferometer with a test object placed behind a ground glass. The 3D temperature profile in a gaseous flame was reconstructed using an Abel transform (see chapter 10). It was noted that using speckle interferometry the temperature profile over the whole field can be visualized simply by observing the double exposure interferogram [268].

Thus, a number of flow configurations were investigated using speckle photography and/or speckle interferometry.

8. Velocity Measurements

Velocity is one of the most important parameter in fluid mechanics due to its cardinal influence on the transport phenomena of heat and mass. Many optical techniques were implemented for velocity measurements. The most highly developed and important among them is *laser Doppler velocimetry*, LDA, see for instance, [270-272].

Most velocity measurements using speckle and speckle-related techniques have at least one common feature with traditional LDV, namely, both approaches use flows seeded by particles and have a great number of common problems in this respect [12]. One may say the same about *particle tracking velocimetry*, PTV, which has a rich bibliography and was developed long before the discovery of speckle methods [273,274]. Recently PTV was cited as the low image density limit of *particle image velocimetry*, PIV, which has a great development as part of 'speckle-related methods'.

PTV is based on the recording of the particle image on the film. This image is moving when the corresponding particle is moving with the flow and the result of the recording is a particle image path on the film which is like a streak. The streak length with a specific time of exposure will give the magnitude of the velocity.

PTV uses a light (laser) sheet approach to illuminate particles in a specific section of the flow under study and provides planar measurement of the velocity field at the chosen instant of time. The most serious problem for PTV until now is automated data extraction from the photographic film with tracer streaks. In semi-automated systems, analysis of the images is performed by manually locating the cross hairs of a precision profile projector on the centres of the initial and final positions of each particle image. Fully computerized image processing systems were developed for this analysis [275,276].

In spite of the PTV technique being very similar to PIV, the first experiments with speckle photography of velocity fields where based not on PTV but on the speckle photography of solid surface deformation.

As early as 1976 Chiang [153] suggested the use of a sheet of coherent illumination as a means of obtaining 3D interior surface displacement vectors by using the illumination of a transparent model in three orthogonal directions through the point of interest and speckle photography of the scattered light from the direction normal to the incident beam.

In the same year, 1976, Barker and Fourney were the first to use the laser sheet and speckle photography method to obtain data about the deformation field of the interior part of a transparent object [154]. They obtained high-quality specklegrams using light scattered from a thin sheet of coherent light passed through a Plexiglas model and showed that these specklegrams could be analysed to yield the interior surface in-plane displacement vectors in the illuminated plane.

In the following year this technique was extended to the measurement of the velocity field of a fluid flow seeded with tracer particles and illuminated by a laser sheet [155-159]. For this technique Dudderar and Simpkins used the term *scattered light speckle* (SLS) metrology; see [160]. Based on measurements of the simultaneous displacements of images of many marked fluid particles, this technique was later named *laser speckle velocimetry* (LSV), *particle image velocimetry* (PIV), *pulsed laser velocimetry* (PLV), or *particle image displacement velocimetry* (PIDV). Of these and a variety of others, the term *particle image velocimetry*, PIV is the most frequently used now.

The wide variety of optical schemes based on the use of the speckle technique for velocity measurements today may be schematically divided into two principal classes. The first class includes the schemes based on *laser sheet* illumination and *static speckle pattern* recording allows one to reconstruct an *instantaneous* two-dimensional velocity map [277-280]. The second one is based on methods where space-time correlation functions are determined by using dynamic speckle pattern recording.

Many attempts have been made to extend such 2D measurements to 3D velocity mapping. The whole three-dimensional field of the flow under consideration can be constructed using these method using plane-by-plane measurements for stationary flows. Modern developments of these methods for studying 3D flows based on the holographic technique are briefly discussed in Section 8.2; see also [281].

8.1 Planar velocity measurements

The principles of laser speckle velocimetry are very simple. The technique involves two important steps. In the first step a selected plane of a seeded flow is illuminated by a light sheet and recorded. The light scattered by the seeding particles in the illuminated plane provides a moving pattern at the recording plane. A double or multiple exposure photograph, called here a specklegram, records this moving pattern.

In the second step the local fluid velocity is derived from the ratio of measured spacing between the images produced by the same tracers, and the time between exposures [61]. Both point-by-point analysis and the spatial filtering technique may be used at this step, which is universal for most speckle methods described throughout the book and discussed in chapter 6.

8.1.1 Light sheet formation techniques

Different types of lasers are now used for light sheet formation. These lasers must provide a sufficient amount of energy for recording the scattered light during a short exposure time. Most frequently a double pulsed ruby laser or a frequency doubled neodim laser is used. Firstly, a laser beam is steered and focused to a small diameter using an inverse telescope lens arrangement. Then, a cylindrical negative lens diverges the focused beam one-dimensionally, creating the light sheet. At a typical laser sheet width of 50 mm, the thickness of the sheet is about of 0.1 mm. This configuration allows one to exposure 35 mm 400 ASA Kodak TMAX film through a 150 mm macro lens (aperture is $f/5.6$ and magnification is 0.5) using 15 mJ/pulse laser light at 0.532 nm wavelength and smoke marker particles [282]. The laser pulse time of about 10 ns and the pulse separation of 50 μs are much smaller than any relevant time scale of the flow field for subsonic flows, thus double exposure photograph truly represents a flowfield at a given instant of time.

High energy is necessary to obtain adequate scattering from marker particles which are very small and/or when a high resolution film is used. Thus for recording 1- 2 μm liquid droplets, and the same Kodak TMAX film, a double-pulsed ruby laser with 2 joule per pulse and time interval between pulses from 1 μs to 10 ms was used in [283]. A 10 joule, Q-switched ruby laser was used for PIV measurements in low-speed aerodynamic applications at Heriot-Watt University [64].

A high repetition rate Cu-vapour laser with an average power of 20 W was used for laser sheet formation in experiments [55]. This laser delivers 2 mJ per pulse with a 50 ns pulse width at a repetition rate 10 000 Hz and provides a time interval between subsequent pulses of 100 μs. The wavelength of radiation was tuned to 0.504 μm. A mechanical (camera) shutter was used here to switch the illumination of the flow on/off. With an exposure time of 1 ms this shutter transmits 10–12 laser pulses. With the shortest possible mechanical shutter time (of about 250 μs) the number of pulses can be decreased to 3. Five–six pulses was found to be optimal for multi-pulsed PIV records.

Continuous wave (CW) lasers have also been used to provide laser sheet illumination [47,49,284,285]. Thus, for studying very low-speed flows at velocities of about 1 mm/s, Meynart used a CW 3 W argon-ion laser, whose radiation was periodically interrupted by using a mechanical shutter providing multi-pulsed laser operation. As the maximum expected velocity increases, shorter pulse times are required to freeze the motion of the seeding particles on to the film. The modulation of the light sheet results in only a small fraction of the available light energy being utilized to illuminate the flow. To avoid this limitation, an alternative technique based on scanning an unexpanded beam through the region of interest within the flow was developed at the University of Edinburgh [285] and is schematically shown in Fig. 8.1.

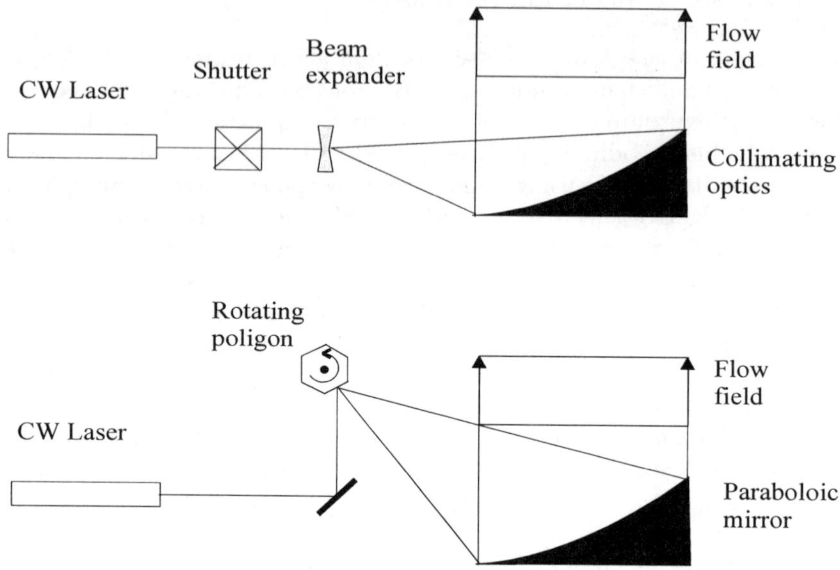

Fig. 8.1. Light sheet formation using CW lasers. Top, modulation by shutter; bottom, a scanning system (from Gray *et al.* [285])

A laser beam was directed here on to a rapidly rotating mirror which has several faces. Mirrors reflect the beam successively through an arc of $4\pi/N$ where N is the number of faces on the mirror. The beam then passes through this arc every $T = 1/FN$ seconds where F is the frequency at which the mirror is rotating. If the polygon mirror deflects the laser beam on to a parabolic mirror then the angular rotation of the beam is translated into a linear velocity scan through the area of interest. This illumination was called *a pseudo light sheet* [285].

In this way each particle within that region of the flow is illuminated for a brief period of time as the laser beam passes and is then re-illuminated when the scanning beam returns to that point of the flow. The exposure at the image plane for such an arrangement is given by

$$E_{\mathrm{S}} = C_{\mathrm{S}} I_{\mathrm{S}} t_{\mathrm{S}} \tag{8.1}$$

where C_{S} is the light loss factor which accounts for the fraction of the light incident on the particle which reaches the image plane of the recording optics, and t_{S} is the time that the laser beam illuminates the particle

$$t_{\mathrm{S}} = \frac{T d_{\mathrm{b}}}{L} \tag{8.2}$$

with L being the horizontal distance through which the beam scans in T seconds, d_b being the diameter of the beam, and T being the time interval between exposures.

As was shown in [285], the time interval between the pulses differs little from T. The motion of the particles between beam passes will modify the distance and therefore the time required for the beam to reach the same particle for the second exposure. The modified time between exposures is given by

$$\delta t = \frac{L - \delta x}{LFN} \tag{8.3}$$

where δx is the horizontal distance travelled by the particle between exposures, $\delta x = v_x \delta t$, and v_x is the averaged horizontal velocity of the particles in time δt. Therefore the expressions used to determine the horizontal and vertical components of velocity from the particle separation, when the particle has been recorded using the scanning beam illumination method, are

$$v_x = \frac{\delta x}{\delta t} = \frac{\delta x F N}{1 - (\delta x / L)} \tag{8.4}$$

and

$$v_y = \frac{\delta y}{\delta t} = \delta y F N \left(\frac{L}{L - \delta x} \right) . \tag{8.5}$$

Typically $\delta x << L$ and so the correction to the measured velocity represents only a small adjustment which can be neglected in most cases.

A simple analysis shows that using the scanning beam the available exposure energy can be increased many fold as compared to the light chopper method. That is, using the scanning beam illumination method, one can generate the same exposure with a smaller laser.

Furthermore, the scanning beam illumination method gives a superior sheet intensity profile. When a cylindrical lens is used to generate a light sheet the Gaussian intensity profile of the laser beam makes a uniform light intensity difficult to achieve. With the scanning beam approach however there is no such problem: an almost perfectly uniform sheet intensity is automatically achieved [285].

A similar scanning method was also developed at Saitama University [286], and called *beam sweep laser velocimetry*. Using this technique with a low-power 15 mW He-Ne laser, Kawahashi and Hosoi provided flow velocity mapping in an area of about 30×50 mm^2 with mean flow velocity of about 20 mm/s.

8.1.2 Particle image and speckle modes

Depending on the tracer concentration, the light scattered by the seeding particles can create two different patterns. This results in two modes of operation of the technique, *particle image velocimetry* (PIV) or *laser speckle velocimetry* (LSV), see [287]. Thus, for low seeding concentration, the pattern consists of resolved 'diffraction limited images' of the particles. When the concentration increases, the images overlap and interfere in the image plane to produce a random 'speckle' pattern.

There are advantages and disadvantages in both modes of operation. In the PIV mode, due to dilute seeding, regions the of flowfield may be left unseeded. In the LSV mode, the slight out-of-plane motion of the scatterers, due to three-dimensional flow between the exposures, results in a different speckle pattern. As a consequence Young's fringes with poor contrast and signal to noise ratio will be produced at corresponding points of the specklegram [61]. This poses a severe restriction in the use of LSV mode for the measurement of velocity in a turbulent field. The fringe quality becomes less dependent on the out-of-plane motion when individual particle images are recorded, i.e. in the PIV mode of operation.

To avoid the image overlap effect, the upper limit for the seeding concentration in the PIV mode, C_{p}, is such that

$$C_{\mathrm{p}} << \frac{M^2}{\Delta z d_{\mathrm{i}}^2} \tag{8.6}$$

where Δz is the thickness of the laser sheet and d_{i} is the image diameter, expressed in term of

$$d_{\mathrm{i}} = \sqrt{d_{\mathrm{p}}^2 + d_{\mathrm{s}}^2} \tag{8.7}$$

with d_{p} the particle diameter and d_{s} the edge spread caused by the limited response of the recording optics.

Considering typical values for $d_{\mathrm{i}} = 0.03$ mm, $\Delta z = 0.5$ mm and M=1, one can obtain $C_{\mathrm{p}} << 2000\ \mathrm{mm}^{-3}$.

The lower end for the seeding concentration can be determined by considering that within the interrogation spot enough particle images must be present. Lourenco and Krothapalli have shown that a single particle image pair is ideal because yields fringes with the best SNR [61].

The most noisy case is two image pairs within an interrogation spot, when a system of crossed Young's fringes is formed, and an appropriate level of SNR can be reached at a number of pairs equal to at least four. This yields for the diameter of the interrogation beam d_{b} equal to 0.5 mm

$$C_{\mathrm{p}} \geq \frac{16}{\Delta z \pi d_{\mathrm{b}}} = 50\ \mathrm{mm}^{-3}. \tag{8.8}$$

From these two constraints, Eqs. (8.6) and (8.8), one can deduce that the optimum value for the concentration is about 100 particles/mm^3, see *e.g.*, [48].

8.1.3 Directional ambiguity

We will briefly discuss the problem of directional ambiguity of the obtained data. Because the order of the first and second exposure images is not recorded, this generally gives two ambiguous speckle displacement vectors of equal magnitude but opposite direction. One of the easiest ways to resolve this ambiguity is to shift the film between two successive exposures as is done in line-of-sight speckle photography. The difference is that in the PIV technique the time interval between two exposures is very short and special efforts are need for such shifting. In this way, Adrian has implemented an image shifting system with a rotating mirror [277]. The system is shown in Fig. 8.2. In this technique, the second image of the speckle pattern is shifted with respect to the first so that all values of displacements are positive.

The position of each object on the film will depend on the mirror angle. By recording the first image with the mirror in one position, and the second image with the mirror rotated, a shift will be introduced. In order to have 'sharp' images of particles during exposures, the exposure time must be essentially less than the characteristic time of the rotation. Image shifting may also be implemented using a birefringent crystal placed before the image plane. Light passing through a birefringent material is displaced for light polarized in one direction, but passes through without displacement for the other polarization. So, to introduce a shift between exposure, the light sheet is polarized horizontally for the first laser pulse and vertically for the second laser pulse [288].

Coupland, Pickering and Halliwell [289,290] proposed solving the ambiguity problem for particle image velocimetry by recording two successive image patterns in such a way that their intensity distributions could be recovered and viewed independently and in known order. For this they have used a holographic approach with two separate reference beam arrangements for each exposure. Even in the theoretically worst case of two particles in the image this technique was able to remove the ambiguity associated with not knowing which distribution was recorded first.

A very elegant way of directional resolution in particle image velocimetry was proposed and implemented recently by Ruck [291]. This approach is based on the temporal modulation of the light intensity illumination. Using a Bragg cell, Ruck formed a laser light intensity profile as vector-like pulses and use this modulated light sheet for particles illumination. This allows one to record the particle trace in a vector-like form, as well. An example of Ruck's particle trajectories record is shown in Fig. 8.3. A very simple mechanical version of the light intensity modulator using a chopper disk with a slit equipped with attenuaters was also realized instead of a Bragg cell.

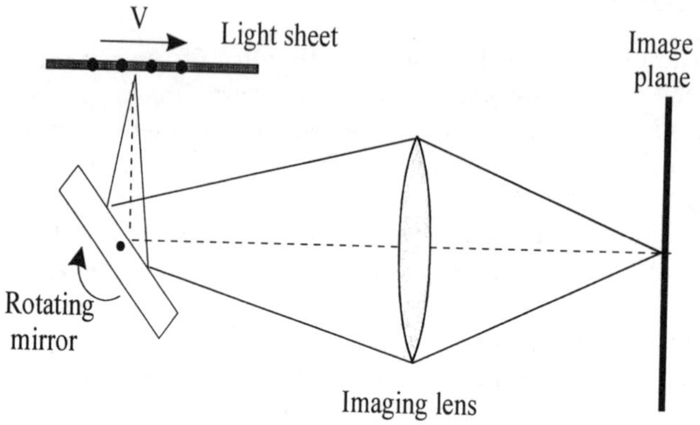

Fig. 8.2. Image shifting using a rotating mirror

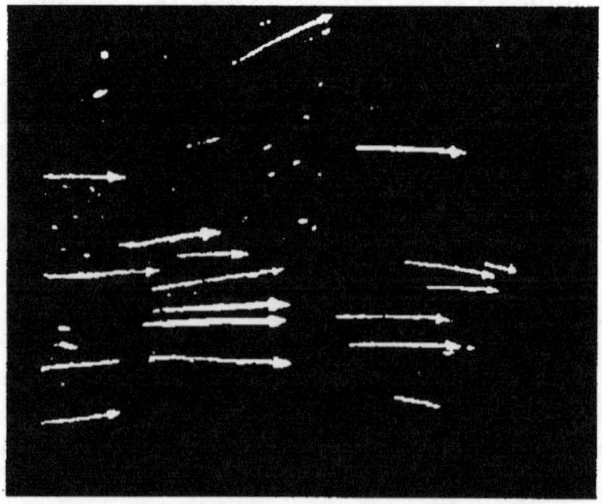

Fig. 8.3. Vector-like particle traces in a turbulent flow, visualized by vector-like illumination pulses (from B.Ruck [291])

There are a number of alternative methods for directional ambiguity resolution described in literature [64,175,277,280,292].

8.1.4 Liquid crystal PIV

Liquid crystals (LCs) have been used in fluid mechanical measurements both in aerodynamics and in water tunnels [293-295]. Temperature changes in such crystals disturb the intermolecular forces and cause a shift in the molecular structure, thus causing a shift in wavelength of the light scattered from the crystals [12]. An extensive literature is available on LCs and their applications [296,297].

Liquid crystal tracers may also be used in PIV for simultaneous determination of temperature (or stresses) and velocity fields [298-302]. Microencapsulated cholesteric LCs 200 μm in diameter were used in [299]. The thermosensitive tracers were illuminated by a halogen lamp. The refracted light was observed by a colour TV camera at right angles to the light sheet. According to the temperature calibration of the liquid crystals, the red, green and blue colours were equal to 29.3, 20.5, and 31.8°C, respectively. Thermal flow stratification in silicon oil was studied using this technique. The cross-correlation calculations using grey levels were used for velocity field determination, whereas colour image processing was used for the reconstruction of the temperature field.

Ozawa *et al.* [301] have used an LC tracer for studying natural convection. The tracer was a powder consisting of microcapsules of about 10–20 μm in diameter filled with a liquid crystal. The LCs were the thermosensitive cholesteric liquid crystals of type 'R 30' produced by Japan Capsular Products Inc., Tokyo. The temperature response range of this LC is also 2.5 K. The colour changes from brown to yellow, green and to blue with increasing fluid temperature. Outside this sensitive range no particular colour can be observed. The response time of the colour change of the LCs is of the order of milliseconds. The density of the LC tracer is slightly higher than that of silicon oil with its values of 900–970 kg/m^3.

Wozniak and Wozniak [302] used encapsulated LC tracers of about 12 μm in diameter (type TCC 1001, BHD Chemical Ltd.). This LC has approximately the same temperature sensitivity as described above.

Dabiri and Gharib have described an automated system for colour image analysis [300]. Their LCs were 150 μm in diameter, reflecting red at 23°C, and reflecting blue at 26°C. The thermal diffusivity of these LCs was 5.8×10^{-8}m^2/s, corresponding to a time response of 0.06 seconds. Xenon short-arc lamp was used as an illumination source. A heated vortex-ring was investigated. It was shown that the errors of the temperature field reconstruction are within 5% of the field.

The obtained results show that LCs can be effectively used in PIV schemes for simultaneous temperature measurements.

8.1.5 Auto- and cross-correlation analysis

The most common form of PIV specklegram treatment is based on correlation processing [137]. The autocorrelation technique described in Section 6.1.5. The technique is based on the analysis of the autocorrelation function of a single (double- or multi-exposed) specklegram.

The autocorrelation function is characterized by a central self-correlation peak plus two or more displacement correlation peaks. The distance from the central peak to either of the displacement correlation peaks is proportional to the average particle displacement within the interrogation area [137].

Analysing two separate images, it is possible to construct the cross-correlation function of these images

$$R_{I_{1-2}}(p,q) = \frac{N \cdot M}{(N-p)\cdot(M-q)} \times \left[\frac{\sum\limits_{n=1}^{N-p}\sum\limits_{m=1}^{M-q} I_1(n,m)\cdot I_2(n+p,m+q)}{\sum\limits_{n=1}^{N}\sum\limits_{m=1}^{M} I_1(n,m)\cdot I_2(n,m)}\right]. \tag{8.9}$$

This *cross-correlation function* (CCF) consists of a single displacement correlation peak. There is no self-correlation peak in this case as there are no identical parts in both interrogation areas. There is no symmetry in the CCF either as the directional ambiguity that arises from recording both exposures on to the same image is removed [137].

The 1D CCF can be determined along any particular direction

$$R_{I1D_{1-2}}(p) = \frac{N}{(N-p)}\left[\frac{\sum\limits_{n=1}^{N-p} I_1(n,m)\cdot I_2(n+p,m)}{\sum\limits_{n=1}^{N} I_1(n,m)\cdot I_2(n,m)}\right]. \tag{8.10}$$

8.1.6 Statistical flowfield analysis

Measurements of turbulence parameters are of great importance for many fluid mechanical applications. The difficulties in measuring turbulence are caused by its nature, namely, turbulent motion is random, fluctuating and three-dimensional flow [303].

Whole flowfield analysis. A high spatial resolution of the velocity measurement can be used both to measure mean flow and turbulence characteristics.

An example of a flowfield in a turbulent jet is shown in Fig. 8.4 [55]. These data were obtained with the help of a Cu-vapour laser as an illumination source.

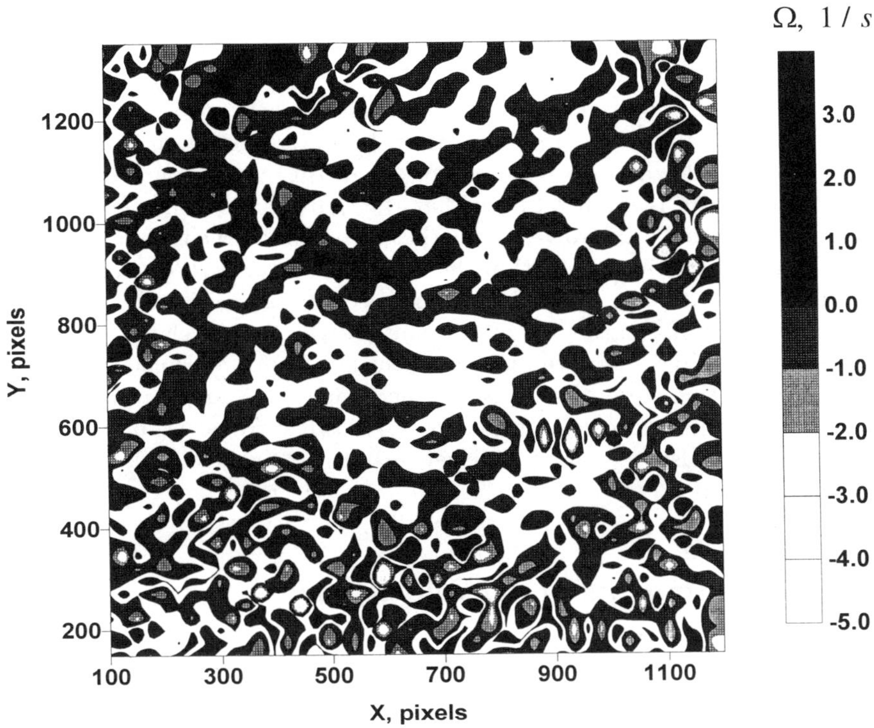

Fig. 8.4. Vorticity field as a result of autocorrelation processing of a multi-exposed PIV image (from [55])

Using automated PIV record treatment, Liu *et al.* [283] have obtained two components of velocity vector at 12 000 points of fully developed turbulent flow simultaneously. Seeding aluminium oxide particles with a diameter of 5 microns were used at a concentration of approximately 35 particles per mm^3. Eleven PIV records, taken at statistically independent times, were interrogated. The instantaneous turbulent velocities were obtained by subtracting the ensemble-averaged velocity. Then, the instantaneous Reynolds stress and spanwise vorticity distribution were obtained. The measurements have shown close agreement with LDA measurements and computer simulation results.

Grant *et al.* [304] have analysed twenty instantaneous PIV records of the same turbulent flow behind a rearward facing step. Each individual PIV record was divided into regular spatial windows. All velocity vectors calculated within a particular window were averaged to produce a mean velocity vector. These regularly positioned velocity vectors were then ensemble averaged using the results from all twenty PIV records. Thus mean and turbulent velocity components were calculated at each spatial window. A confidence analysis has been outlined and applied to PIV sampling to demonstrate a method of assessing the reliability of the mean velocity and turbulence measurements obtained [304].

Local turbulence analysis. It has been assumed that the constant particle (speckle) displacements there are within the interrogation area. Small-scale turbulence could disturb this assumption. In this case the auto- (or cross-) correlation peak is modified such that the 'pure' particle image correlation function has to be convolved with the probability density function of the displacement fluctuations within the interrogation area [175]. With growing standard deviation it spreads and decreases in size [175]. This opens the way to obtain information about flow statistics by the correlation peak shape analysis [305,306]. As shown by Hinsch, the non-uniform displacement within the interrogation region leads also to decreasing the Young's fringe visibility in point-by-point specklegram analysis [175]. The fringe visibility analysis, therefore, has also been used for local turbulence diagnostics in a number of papers [307-310].

8.2 Measurements of 3D velocity fields

8.2.1 Stereoscopic particle image velocimetry

As demonstrated in the previous section, the PIV technique provides an instantaneous map of the two in-plane components of the velocity field. The plane is selected by illumination of the flow with a laser light sheet. If the flow is three-dimensional, the out-of-plane component is a source of error[1] in the measured velocity [311]. It was found that this error was directly related to the distance from the point to the axis of the recording system. On the other hand, stereoscopic imaging is a conventional three-dimensional mapping approach for particle-tracking velocimetry in which a particle-seeded flow is viewed from different angles [312,313]. Therefore, the stereoscopic methods, in which the flow is viewed from two different positions, were effectively used in PIV as well as [314-317].

Two different stereoscopic approaches are described in the literature. In *the angular approach*, the optical axis of the two recording cameras is no longer perpendicular to the flow but makes a certain angle with the illuminated sheet. In the *translation method*, the stereoscopic effects are directly related to the distance between the optical axes of the cameras, which are now perpendicular to the illuminated sheet. The theory of *translational stereoscopic* PIV is given by Arroyo and Greated [315].

[1] There attempts to extract the third velocity component from the 2D map by applying the 3D continuity equation, see e.g. [51]

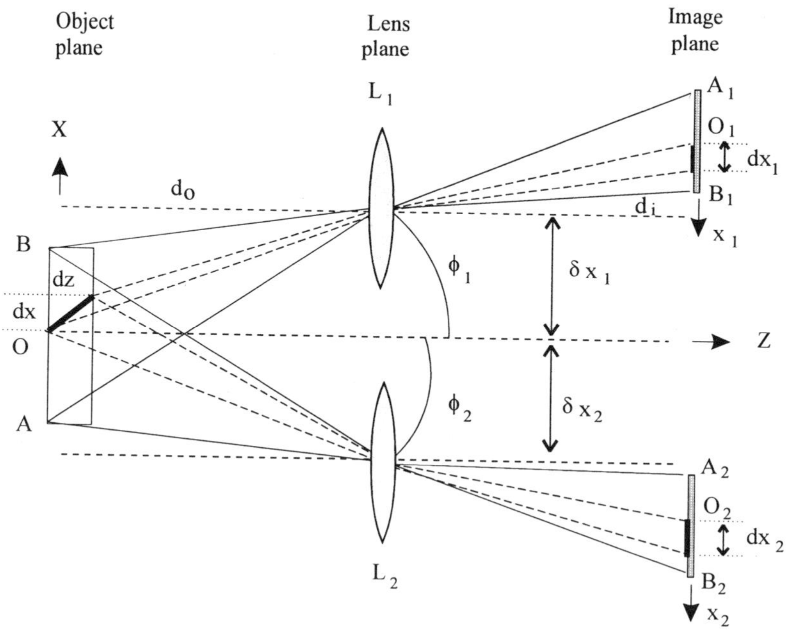

Fig. 8.5. Stereoscopic translational method (from Arroyo and Greated [315])

Following this publication, consider two cameras with their optical axes normal to the laser sheet at distances δx_1 and δx_2 from the centre of the object, see Fig. 8.5.

The coordinates of the two lens centres will be $(\delta x_{\mathrm{i}}, 0, d_{\mathrm{O}})$, where d_{O} is the object distance. The coordinates $(x_{\mathrm{j}}, y_{\mathrm{j}}), j = 1, 2$, of a point (x, y, z) in the laser sheet are

$$x_{\mathrm{j}} = \frac{d_{\mathrm{i}}}{d_{\mathrm{O}} - z}(x - \delta x_{\mathrm{j}}), \tag{8.11}$$

$$y_{\mathrm{j}} = \frac{d_{\mathrm{i}}}{d_{\mathrm{O}} - z} y, \qquad j = 1, 2$$

where d_{i} is the image distance, and the origin for $(x_{\mathrm{j}}, x_{\mathrm{j}})$ is the centre of each lens.

If the particle moves to the point $(x+\mathrm{d}x, y+\mathrm{d}y, z+\mathrm{d}z)$ its new coordinates will be

$$x_{\mathrm{j}} + \mathrm{d}x_{\mathrm{j}} = \frac{d_{\mathrm{i}}}{d_{\mathrm{O}} - (z + \mathrm{d}z)}(x + \mathrm{d}x - \delta x_{\mathrm{j}}), \tag{8.12}$$

$$y_{\mathrm{j}} + \mathrm{d}y_{\mathrm{j}} = \frac{d_{\mathrm{i}}}{d_{\mathrm{O}} - (z + \mathrm{d}z)}(y + \mathrm{d}y), \qquad j = 1, 2.$$

Assuming that $z, \mathrm{d}z << d_0$, the displacement $(\mathrm{d}x_\mathrm{j}, \mathrm{d}y_\mathrm{j})$ can be expressed as

$$\mathrm{d}x_\mathrm{j} = M\left(\mathrm{d}x + \frac{\mathrm{d}z}{d_0}(x - \delta x_\mathrm{j})\right), \tag{8.13}$$

$$\mathrm{d}y_\mathrm{j} = M\left(\mathrm{d}y + \frac{\mathrm{d}z}{d_0}y\right), \qquad \mathrm{j} = 1, 2$$

with M being the magnification of the recording system, $M = d_\mathrm{i}/d_0$.

Finally, the displacement vector $\mathbf{dr} = (\mathrm{d}x, \mathrm{d}y, \mathrm{d}z)$ can be obtained as

$$\mathrm{d}z = d_0 \frac{\mathrm{d}x_2 - \mathrm{d}x_1}{M(\delta x_1 - \delta x_2)}, \tag{8.14}$$

$$\mathrm{d}x = \frac{\mathrm{d}x_\mathrm{j}}{M} + \frac{\mathrm{d}z}{d_0}\frac{(x_\mathrm{j} - x_{j0} - M\delta x_\mathrm{j})}{M},$$

$$\mathrm{d}y = \frac{\mathrm{d}y_\mathrm{j}}{M} + \frac{\mathrm{d}z}{d_0}\frac{(y_\mathrm{j} - y_{j0})}{M}, \qquad \mathrm{j} = 1, 2.$$

Using these expressions, the 3D vector of velocity is determined as

$$\mathbf{v} = \frac{\mathbf{dr}}{T}, \tag{8.15}$$

where T is the time interval between two consecutive pulses.

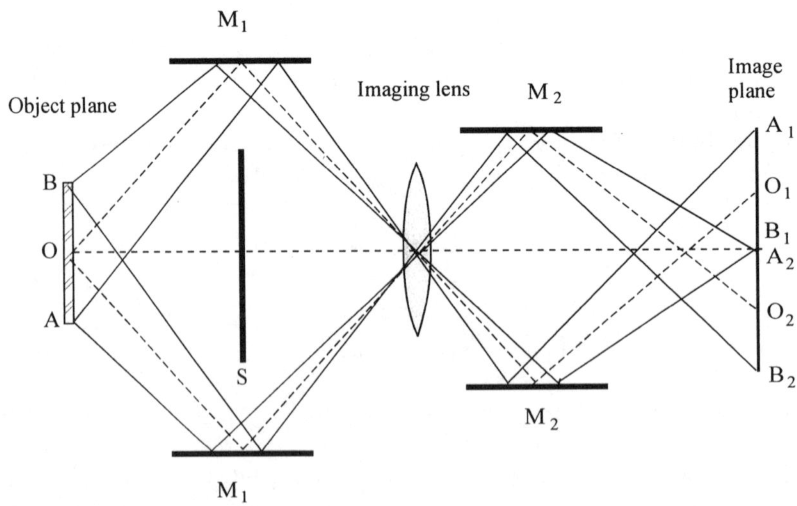

Fig. 8.6. Four-mirror optical system for the stereoscopic translational method (from Arroyo and Greated [315])

Arroyo and Greated performed stereoscopic PIV using the four-mirror optical system shown in Fig. 8.6 [315]. The first two mirrors of the system must be perfectly perpendicular to the illuminated sheet in order to have the optical axis of the equivalent two lenses perpendicular to the sheet. This optical system will also give $\delta x_1 = -\delta x_2 = \delta x$ if two mirrors are centred with respect to the object and the lens in order to have the light collected from the object by both mirrors passing just through the central part of the lens.

The purpose of the second pair of mirrors is to redirect the light so that the two images are formed on the back of the camera, each one in a different half of the film. They must be perfectly perpendicular to the film.

The set-up can be easily converted into standard PIV by removing the blockage S and covering the first two mirrors with an opaque object. The technique does not need any calibration procedure, apart from measuring the value of magnification, as in *single image* PIV. Thus, using this optical set-up, two *stereoscopic* images are taken simultaneously with the same camera. Analysing these images with the usual techniques, the three velocity components are obtained by combining the in-plane measurements from both images.

The system was tested on measurements of acoustic streaming flow in a rectangular tube [315], and in multi-phase flows [318]. Analysis shows that the error in the out-of-plane component is about three times greater than the error in the in-plane components, which remains the same as in the *single image* 2D PIV system.

Prasad and Adrian [317] have used a twin-camera stereoscopic system. They have tested the accuracy and resolution the system. Translation tests have yielded errors (rms) of 0.2% of full-scale for the in-plane displacement, and 0.8% of full-scale for the out-of-plane component.

Grant *et al.* [319] have described on-line stereoscopic system. Cenedese and Paglialunga have described a new technique for the determination of the third velocity component by using of two parallel light sheets of different colours in a single image PIV scheme [320].

8.2.2 Holographic particle image velocimetry

Holographic particle recording. Holography is a well-established means of particle field diagnostics [189,321]. The holographic method of particle recording in PIV provides a three-dimensional velocity measurement; see e.g. [281]. The processing methods for holographic PIV records are basically the same as for photographic records [217]. A major problem in holographic PIV is the discrimination of the particle positions in the direction of the camera. Studies of the depth discrimination problem have been made by Moratis and Buchhave [322], by Gray and Greated [323], and by Meng and Hussain [324]. It was demonstrated that three-dimensional directionally resolved displacements can be measured from double exposure particle field holograms. The accuracy and resolution of measurements in the axial direction appear to be poorer than in-plane measurements by more than an order of magnitude [323]. Nevertheless, using holographic PIV 10,824 velocity vectors in a 3D

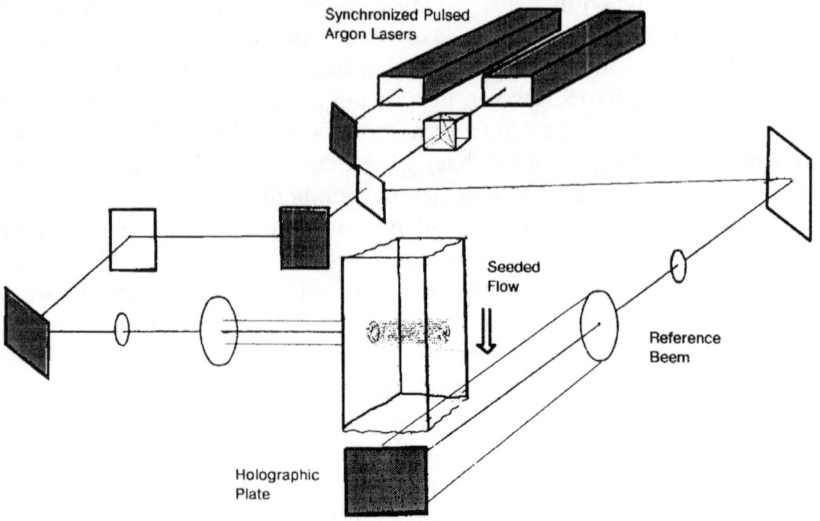

Fig. 8.7. Holographic installation for PIV recordings at the University of Edinburgh (from Gray and Greated [323])

vortics were reconstructed from one hologram in a 3D space domain with spatial resolution equal 1 mm in all dimensions [324]. Zhang *et al.* [325] using HPIV with high pass filters and automated image acquisition and processing system, have reconstructed 97×97 × 87 vectors in the sample volume with dimensions 46.60×46.60 × 42.25 mm^3. Thus, 818,583 velocity vectors were reconstructed using one holographic plate.

In spite of the fact that the holographic PIV technique appeared not very long ago, there are already many publications devoted to this promising direction of velocimetry [325-331]. The local particle displacements are determined by using stereoscopic viewing for depth perception [326,329]. Two cameras viewing the same image region from two different angles can acquire stereo images, from which the 3D particle displacement can be calculated.

A typical holographic installation for PIV velocimetry is shown in Fig. 8.7. An off-axis reference beam system is used so that the virtual image of the particle field is viewable separately from the real image and the direct transmitted reference beam [323]. A double exposure hologram is made, recording particle positions at two instants of time on to the same holographic film.

Synchronized pulsed argon lasers deliver two short pulses (10 ns duration) of 0.1 J each at a wavelength of 0.514 μm and a pulse separation in time of 100 μs.

The *'light-in-flight' holography* is also used to select a single sheet in depth during holographic PIV records [175,332]. Lasers with relatively short coherence length are needed for this interesting experimental scheme. A Q-switched ruby laser could be used for such measurements. With a single intracavity etalon its coherence length can be adjusted to measure about 10 cm [175].

Holographic interferometry. Holographic interferometry is used for simultaneous measurements of the three components of the velocity field in the whole 2D plane. The method is based on the analysis of several interferograms with different sensitivity vectors, obtained from the same hologram by varying the viewing direction [333,334]. A phase-shift holographic interferometer with digital hologram recording using a CCD camera was deleloped by Skarman *et al.* [335].

Thus, three-dimensional PIV becomes feasible with the use of holographic technique.

8.3 Dynamic speckle method

The time-varying intensity of speckles formed in the diffraction and image fields by scattering coherent light from a moving diffuse object carries information about the object motion [35,336,337].

Applications of *dynamic speckles* to speckle metrology have been studied extensively by many researchers. An important application is velocity measurement. Special attention is paid to velocity measurements of bio-objects, like *in vivo* blood flux diagnostics.

Dynamic speckle velocimetry is based on the spatio-temporal correlation analysis of the speckle intensity fluctuations at the receiver. Figure 8.8 shows an arrangement for the velocity diagnostics by the dynamic speckle method [336].

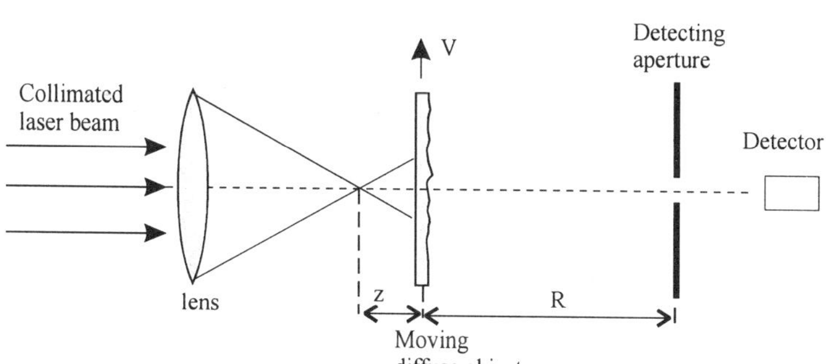

Fig. 8.8. Illustration of dynamic speckle method

In this figure, the time-varying speckles from the moving diffuse object under illumination by a Gaussian beam are formed in the Fresnel diffraction field, see Section 3.3.1. At the vector coordinate $\mathbf{r}$ representing a point at the object plane, the amplitude distribution of the light illuminating the object surface situated at a distance z away from the waist position is given by [336]

$$U(\mathbf{r}) = \frac{w_0}{w}\exp(\mathrm{i}\frac{2\pi z}{\lambda}) \cdot \exp(-\frac{|\mathbf{r}|^2}{w^2}) \cdot \exp(\mathrm{i}\frac{\pi|\mathbf{r}|^2}{\lambda\rho}), \tag{8.16}$$

where w_0 is the width of the beam waist. The parameters w and ρ are the z-dependent width and wavefront curvature of the beam, respectively, and are given by

$$w(z) = w_0(1 + z^2/a^2)^{1/2} \tag{8.17}$$

and

$$\rho(z) = z(1 + a^2/z^2), \tag{8.18}$$

where $a = \pi w_0^2/\lambda$. Takai *et al.* [336] have shown that the normalized temporal autocorrelarion function of time-varying speckle intensity fluctuations $\Delta I(t) = I(t) - < I(t) >$ detected at a certain point in the Fresnel diffraction field is given by

$$\mathcal{R}_{\Delta I}(t) = \exp(-\tau^2/\tau_{\mathrm{C}}^2). \tag{8.19}$$

with $\tau_{\mathrm{C}} = K/|v|$ where the proportionality constant is related to the parameters of the optical configuration by

$$K = (\frac{1}{w^2} + \frac{\pi^2 w^2 \sigma^2}{\lambda^2 R^2})^{-1/2} \tag{8.20}$$

where $\sigma = R/\rho + 1$ with R being the distance from the moving object to the recording plane.

The normalized power spectral density function of the speckle intensity fluctuations is then obtaining using the Wiener–Khinchine theorem and is given by

$$\mathcal{W}_{\Delta I}(\omega) = \exp(-\tau_{\mathrm{C}}^2\omega^2/4). \tag{8.21}$$

Thus, using the relations (8.20) and/or (8.21) the velocity may be measured by analysis of the intensity autocorrelation function or power spectral density function.

A related speckle dynamic method was used by Fujii et al. [338] for monitoring subskin blood flow. They showed the usefulness of the method for clinical purposes. With the overall analysis of a time-varying laser speckle, stable data on the skin blood flow velocity are easily obtainable from this method even for the smallest body capillaries, like in retinal blood flow [339].

An alternative technique, *the single exposure speckle photography method*, was also developed for quantitative velocity distribution determination by mapping the variation of speckle contrast [104,107,108]. In particular, this technique was applied to the visualization of blood flow in the human retina

[104]. The method is based on the fact that in finite-time exposure speckle photograph, the images of the speckles will be blurred in an area where flow is occurring. The speckle contrast in this area will be decreased to an extent which will depend on the flow velocity and the exposure time. For random motion of the scatterers, the speckle pattern will decorrelate in time but will not drift, and the spatial variance in the time-averaged speckle pattern is given by [104],

$$\sigma_{\mathrm{S}}^2 = \frac{1}{T}\int\limits_0^T C_{\mathrm{t}}^2(\tau)\,\mathrm{d}\tau, \tag{8.22}$$

with T being the exposure time.

The function $C_{\mathrm{t}}^2(\tau)$ is the autocovariance of the temporal fluctuations in the intensity of a single speckle

$$C_{\mathrm{t}}^2(\tau) =< [I(t)- < I > |_{\mathrm{t}}][I(t+\tau)- < I > |_{\mathrm{t}}] > |_{\mathrm{t}} \tag{8.23}$$

where $< \ldots > |_{\mathrm{t}}$ means time average. Using this relation and assuming a simple negative exponential dependence for the speckle field correlation function, Fercher and Briers have obtained a relationship between the speckle contrast and the speckle field 'correlation time', see [104]

$$\frac{\sigma_{\mathrm{S}}}{< I >} = \left[\frac{\tau_{\mathrm{C}}}{2T}\left\{1 - \exp\left(\frac{\tau_{\mathrm{C}}}{2T}\right)\right\}\right]^{\frac{1}{2}}. \tag{8.24}$$

The time average here was replaced by the ensemble average using ergodicity.

In an area of no flow the speckle pattern will remain 'sharp' with the initial value of the contrast.

This technique requires calibration, and, as was shown by Briers [107] in the case of the human retina, an exposure time of 1 ms with a helium-neon laser ($\lambda = 0.633$ nm) should allow the mapping of blood velocities between 0.02 and 5 mm s^{-1}. This work was further developed in [108] to quasi-real time recording by using CCD cameras.

9. Measurements of Turbulence Parameters: Scalar Fields

In turbulent flow both fluid velocity and scalar quantities, like temperature, density, or concentration, may be separated into mean and fluctuating ensemble average components. The turbulence of the scalar fields is superimposed to the turbulent velocity field. Both velocity and scalar statistics are necessary for the description of turbulent heat and mass transfer. Diagnostics of the turbulent parameters based on the velocity statistics are discussed in the previous chapter. The methods are based on the statistical analysis of the particle seeded into the flow displacement under light sheet illumination. This chapter is devoted to the scalar flowfield statistical analysis using line-of-sight flow probing. An obvious advantage of such flow probing is the absence of the necessity to introduce into the flow the seeded particles. A disadvantage of these methods is a spatial data averaging along an optical path. A way to overcome this disadvantage, multi-angular flow probing with subsequent local parameter reconstruction with the help of computerized tomography, will be discussed in the next chapter.

The scalar is said to be passive if its behaviour has no appreciable effect on the dynamics of turbulence. It is in scalar fields that there has been a growing interest in turbulence studies during the past few years, e.g. in connection with combustion processes or pollution problems [340-343].

A great number of experimental results have given evidence of the validity of the concept of passive scalar fields. Many of these experiments were performed in the flow downstream of a turbulence generating grid and with temperature as the scalar quantity; see *e.g.* [244].

Unlike light sheet illumination, when speckle photography provides measurements of the velocity vectors, *line-of-sight speckle photography* measures the mean and fluctuating parts of the refractive index field. For isobaric turbulent flows, the mean and fluctuating parts of the temperature distribution can be evaluated using these data. Here we will discuss *line-of-sight* measurements only. This type of measurement can provide velocity data, as well. Thus, Grandke has shown that by means of the laser shadow technique the mean flow velocity and the turbulence level can be obtained from the spatio-temporal correlation function of a shadowgraph pattern [345,346]. This approach, like the more general approach called *image correlation velocimetry* [133,134], is based on the assumption that no structures exist in the flow which do not travel at the mean velocity (e.g. shocks or Mach waves) [346]. Similar methods are described above in sections devoted to dynamic speckle methods (Section 8.3), and specklegram evaluation by cross-correlation analysis (Section 8.1). So, only scalar turbulent fields, reconstructed with the help of refractive index distributions, will be analysed here.

Two main approaches to turbulence parameter reconstruction by speckle photography will be described.

The first one is based on *instantaneous* density or temperature field recording on very detailed grids. Using the high density of the obtained data, spatial correlations, scales, and energy spectrum are then evaluated by flow-field analysis.

The second approach is based on the recording of *temporal averaged data.* Each interrogation point of the specklegram may be analysed independently from others and provides information about turbulence intensity and anisotropy with a higher spatial resolution. Analysis of the whole specklegram of this type allows one also to evaluate spatial correlations, scales, and the energy spectrum of the turbulence.

9.1 Refraction in a turbulent field

Turbulence affects the propagation of a beam of light through the medium under investigation by way of variations in the refractive index. These variations can be caused by concentration fluctuations in a mixing zone of components with different refractive indexes. For one-component flows the refractive index variations are caused by density fluctuations, or, in the case when the pressure variations are small, by temperature fluctuations.

9.1.1 Statistics of the refractive index

The refractive index in a turbulent field may be decomposed into the mean and fluctuating components

$$n = \langle n \rangle + n' \qquad (9.1)$$
$$\langle n' \rangle = 0.$$

The averaged values can be introduced by time average, space average or ensemble average, see [303]. The turbulence is often assumed to have an additional property, the property of ergodicity, in which the time averages, the space averages and ensemble averages are identical [347].

Probability density function. By introducing a *probability density function*, $\mathcal{P}(n)$, the ensemble average may be expressed as *moments of the probability density function*

$$<n>=\int_{-\infty}^{\infty} n\mathcal{P}(n)\,\mathrm{d}n, \tag{9.2}$$

$$<n^2>=\int_{-\infty}^{\infty} n^2\mathcal{P}(n)\,\mathrm{d}n, \tag{9.3}$$

and so on.

The second central moment, or *the variance*, of the turbulent flow parameter is a measure of the intensity of the turbulence.

$$<(n')^2>=\int_{-\infty}^{\infty} (n')^2\mathcal{P}(n)\,\mathrm{d}n. \tag{9.4}$$

The square root of the variance, or *standard deviation*, of the random variable can be found as follows

$$\sigma_n=\sqrt{<(n)^2>}. \tag{9.5}$$

As $\sigma \to 0$, the random fluctuations vanish and the variable under consideration tends to a completely determined value, namely, its average [348].

Correlation functions. The *autocorrelation function* of the refractive index distribution, also called *covariance function*, can be introduced as follows

$$\mathbf{R}_n(\mathbf{r_1},\mathbf{r_2})=<n'(\mathbf{r_1})\cdot n'(\mathbf{r_2})> \tag{9.6}$$

This definition can be rewritten in the form

$$\mathbf{R}_n(\mathbf{r},\boldsymbol{\Delta}\mathbf{r})=<n'(\mathbf{r})\cdot n'(\mathbf{r}+\Delta\mathbf{r})> \tag{9.7}$$

where $\mathbf{r}=\mathbf{r_1}$ and $\mathbf{r_2}=\mathbf{r_1}+\Delta\mathbf{r}$.

It is convenient to use also the dimensionless autocorrelation coefficient:

$$\mathbf{Q}_n(\mathbf{r},\boldsymbol{\Delta}\mathbf{r})=\frac{\mathbf{R}_n}{\sqrt{<[n'(\mathbf{r})]^2>\cdot<[n'(\mathbf{r}+\Delta\mathbf{r})]^2>}} \tag{9.8}$$

For homogeneous and isotropic turbulence these values are scalar and depend on the scalar argument $|\Delta\mathbf{r}|=\Delta r$.

Using an expansion of the autocorrelation coefficient in a Taylor series in the vicinity of $\Delta r=0$,

$$Q(\Delta r)=1+\frac{(\Delta r)^2}{2!}\cdot\frac{\partial^2 Q}{\partial(\Delta r)^2}+\frac{(\Delta r)^4}{4!}\cdot\frac{\partial^4 Q}{\partial(\Delta r)^4}+\dots \tag{9.9}$$

and saving only quadratic terms in this expansion, we can introduce the correlation length, λ_C

$$\frac{1}{\lambda_C^2} = -\frac{1}{2} \cdot \frac{\partial^2 Q}{\partial(\Delta r)^2} \tag{9.10}$$

so that

$$Q(\Delta r) \simeq 1 - \frac{(\Delta r)^2}{\lambda_C^2}. \tag{9.11}$$

This correlation length is also called the dissipation lengthscale, and may also be considered a measure of the average dimension of the smallest turbulent eddies [303].

It is clear that for non-isotropic turbulence, the three scales for the correlation function must be introduced for all three orthogonal directions.

It is possible to introduce also one another length scale, namely Λ_C, so that

$$\Lambda_C = \int_0^\infty Q(\Delta r)\,\mathrm{d}(\Delta r). \tag{9.12}$$

Using this parameter, the correlation coefficient can also be represented in the form

$$Q(\Delta r) = \exp(-\frac{\Delta r}{\Lambda_C}). \tag{9.13}$$

The value Λ is called the *integral*, or *macroscale* of turbulence.

Structural functions. In addition to the correlation functions it is convenient to use structural functions also. Such structural functions were introduced by A.N. Kolmogorov for velocity fields [349] and by A.M. Obuchov for temperature fields [350] as measures of the strength of the fluctuations.

For the refractive index field in a *locally homogeneous turbulence* this function is

$$\mathbf{D}_n(\mathbf{r}, \mathbf{\Delta r}) = < [n'(\mathbf{r}) - n'(\mathbf{r} + \Delta\mathbf{r})]^2 > \tag{9.14}$$

and for the temperature variation the function is

$$\mathbf{D}_T(\mathbf{r}, \mathbf{\Delta r}) = < [T'(\mathbf{r}) - T'(\mathbf{r} + \Delta\mathbf{r})]^2 > . \tag{9.15}$$

In contrast to the correlation function the structural functions is not sensitive to 'slow' variations of the field under study and shows the microscale structure of the field with spatial frequencies less than Λ_C, the turbulence macroscale. The correlation function according to its definition is sensitive to variations of the field with all spatial frequencies.

The structural function can be expressed through correlation functions and vice versa

$$\mathbf{D}(\mathbf{r}, \mathbf{\Delta r}) = 2 \cdot [\mathbf{R}(\mathbf{r}, 0) - \mathbf{R}(\mathbf{r}, \mathbf{\Delta r})] \tag{9.16}$$

and

$$\mathbf{R}(\mathbf{r}, \boldsymbol{\Delta}\mathbf{r}) = \frac{1}{2} \cdot [\mathbf{D}(\mathbf{r}, \infty) - \mathbf{R}(\mathbf{r}, \boldsymbol{\Delta}\mathbf{r})]. \tag{9.17}$$

For *homogeneous and isotropic turbulence* these values are scalar and depend on the scalar argument $|\boldsymbol{\Delta}\mathbf{r}| = \Delta r$. As mentioned above the structural function describes the variation of the field under study within a small region with $\Delta r \leq L$ where L is the macroscale of the turbulence. For *locally homogeneous turbulence* the structural function can have the same form at different points where the intensity of the turbulence is different. For such cases the structural function can be represented in the form

$$\mathbf{D}(\mathbf{r}, \boldsymbol{\Delta}\mathbf{r}) = \mathbf{C}^2(\mathbf{r}) \cdot D_0(\boldsymbol{\Delta}\mathbf{r}). \tag{9.18}$$

Here the structural constant, $\mathbf{C}(\mathbf{r})$, describes the variation of turbulence intensity from point to point.

Spectral distributions. The *spectrum* or the spectral energy density of the refractive index distribution, $\Phi_n(k)$, may be introduced as the three-dimensional Fourier transform of the correlation function

$$\Phi_n(k) = \frac{1}{(2\pi)^2 k} \int_0^\infty \Delta r \cdot \mathbf{R}_n(\Delta r) \cdot \sin(k\Delta r) \cdot \mathrm{d}\Delta r \tag{9.19}$$

with

$$\mathbf{R}_n(\Delta r) = \frac{4\pi}{\Delta r} \int_0^\infty k \cdot \sin(k\Delta r) \cdot \Phi_n(k)\, \mathrm{d}k \tag{9.20}$$

with k being the wavenumber. In the general case, a vector value of $\mathbf{k}$ must be used. For isotropic fields we can use $k = |\mathbf{k}|$.

The structural function can also be expressed through the spectrum

$$\mathbf{D}(\Delta r) = 8\pi \int_0^\infty (1 - \frac{\sin k\Delta r}{k\Delta r}) \cdot \Phi(k) \cdot k^2\, \mathrm{d}k \tag{9.21}$$

and

$$\Phi(k) = \frac{1}{4\pi^2 k^2} \int_0^\infty \mathbf{D}(\Delta r)[\sin(k\Delta r) - k\Delta r \cdot \cos(k\Delta r)]\, \mathrm{d}\Delta r. \tag{9.22}$$

For isotropic homogeneous fields the structural function can be treated as one-dimensional and can be presented in the form of a one-dimensional spectral distribution

$$\mathbf{D}(\Delta r) = 2 \int_{-\infty}^{\infty} [1 - \cos k_1 \Delta r] \cdot \Phi(k_1) \, \mathrm{d}k_1 \tag{9.23}$$

and

$$\Phi(k_1) = \frac{1}{2\pi k_1} \int_{0}^{\infty} \mathbf{D}'(\Delta r) \cdot \sin(k_1 \Delta r) \, \mathrm{d}\Delta r \tag{9.24}$$

with k_1 being the one-dimensional wavenumber.

Assuming the Kolmogorov model of turbulence, the structural function and the spectrum can be approximated by [28]

$$D_n(r) = C_n^2 r^{2/3} \tag{9.25}$$

and

$$\Phi_{\mathrm{n}}(k) = 0.033 C_{\mathrm{n}}^2 \exp(k/k_{\mathrm{m}}). \tag{9.26}$$

9.1.2 Ray tracing through a turbulent field

To analyse the possibilities of speckle photography for turbulence study a direct numerical simulation of the optical scheme was performed. Under the assumption of geometrical optics the ray paths through the three-dimensional turbulent field are computed from the ray equation,

$$\frac{\mathrm{d}}{\mathrm{d}s}[n(\mathbf{r})\frac{\mathrm{d}\mathbf{r}}{\mathrm{d}s}] = \nabla n(\mathbf{r}), \tag{9.27}$$

where $\mathbf{r} = x\mathbf{i} + y\mathbf{j} + z\mathbf{k}$ is the position vector of a point on the ray and $\mathrm{d}s$ is an element of the arc length along the ray.

As was shown by Tatarski, [28], this ray equation is valid when the conditions

$$\lambda << l_{\mathrm{t}}, \qquad L_1 << \frac{l_{\mathrm{t}}^2}{\lambda} \tag{9.28}$$

are satisfied. Here λ is the wavelength, l_{t} is the turbulence microscale, and L_1 is the propagation distance. Thus for laboratory turbulence modelling, when L_1 is about 0.1 m, the turbulence microscale can be about 1 mm. These values are in good accordance with the spatial resolution of speckle photography and with the finest possible grids for direct numerical simulation of turbulence. The same approach is possible to apply for describing light propagation through a random medium [351] and for acoustic wave tracing [352]. For greater propagation distances, Rhutov's method which includes diffraction effects must be applied [27,353].

The numerical procedure of solving this ray equation is described above in Section 7.1.2. It allows one to compute the ray positions and the slope for a small increment in path length starting from the initial cross-section where the rays are parallel to the z-axis.

The 3D index of refraction field is prescribed on a $64 \times 64 \times 64$ mesh grid using data of a direct numerical simulation of turbulent shear flows obtained by Dr. T.Gerz, [354].

Figure 9.1 contains an example of one of the cross-sections of the isotropic turbulent field obtained in [354] by integration of the exact time-dependent three-dimensional Navier-Stokes and temperature equations in a cubic domain and in time.

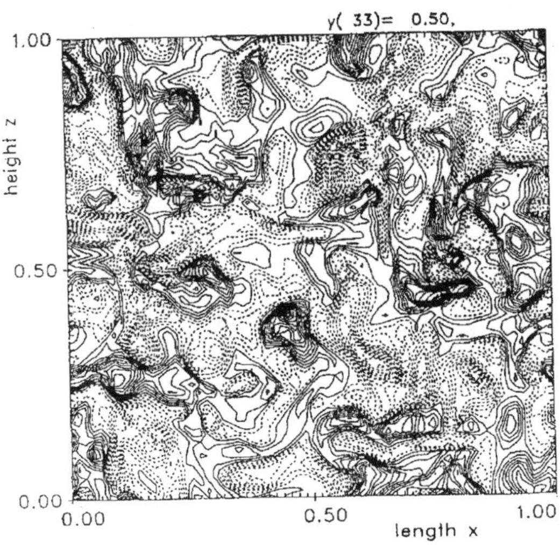

Fig. 9.1. Isolines of temperature fluctuations in one of the cross-sections of an isotropic turbulent field (from Gerz [354])

The ray paths of an initially plane wave which propagates through the computer-generated turbulent field are shown in Fig. 9.2. For high-temperature gas flows at moderate Reynolds numbers the rays propagate along nearly straight lines and typical angles of refraction on turbulent density fluctuations are of the order of 10^{-5}–10^{-3}. Such values exactly correspond to the dynamic range of speckle photography, an optical technique sensitive to the refraction angles of the light passed through the flowfield under study.

For vizualisation of the ray paths in Fig. 9.2 the absolute values of density fluctuations in the turbulent field were artificially increased 100 times. It is possible to see that the first caustics appear after 3-4 correlation lengths for the density field. The same conclusion was made in [351] after studying ray tracing through a computer-generated random medium.

At the exit cross-section the vector values of speckle shifts for different optical schemes were calculated. The correlation functions of the deflection

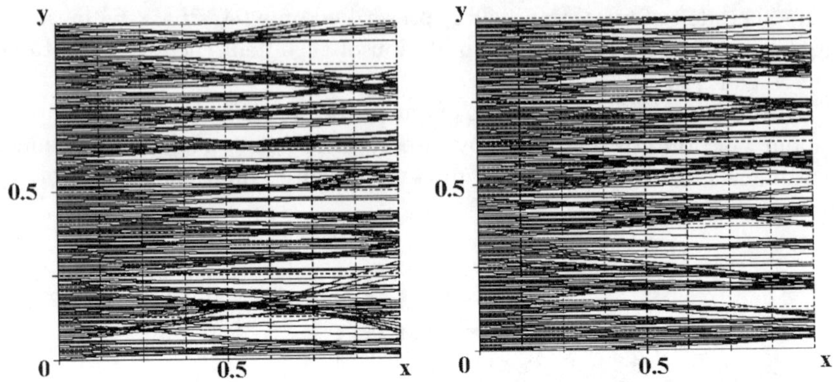

Fig. 9.2. Results of numerical simulation of ray tracing through the turbulent field

angle field were constructed and compared with density correlation functions in the turbulence field, see the next section.

9.2 Spatial correlation analysis by speckle photography

9.2.1 Shadowgraphy of turbulent flows

Spatial correlation analysis of a specklegram is essentially the same as for shadowgraphy. This image analysis was initiated by the publication of Uberoi and Kovasznay [355] as early as 1955. In this paper the optical response equation for the autocorrelation of light intensity fluctuations of a plane optical wave traversing a turbulent field was derived.

Uberoi and Kovasznay based their consideration upon the Weyl approximation which assumes the use of geometric optics and considers the turbulent layer to be thin compared to its distance to the shadowgraph plate. Taylor [356] extended such considerations to an arbitrary thickness of the turbulent layer. He used the Born approximation for the solution of the wave equation, based on iteration. It was shown that the first iteration gives an appropriate results for small density fluctuations, $\frac{\Delta\rho}{\rho_0} < 0.02$. The shadowgraph method as well as the Schlieren techniques have been extensively employed in turbulence studies [367-360].

9.2.2 Reconstruction of correlation functions

The high signal density that can be achieved using automated specklegram analysis makes the speckle photographic method particularly attractive for the analysis of turbulent flows. However, the spatial resolution of speckle photography isn't infinite and amounts to about of 0.2–0.5 mm. This means that refractive index fluctuations having higher spatial frequencies can't be recorded by speckle photography and make a contribution to the decorrelation effects decreasing the Young's fringes contrast under the treatment of a specklegram obtained for turbulent flow.

In the early days of the speckle technique for fluid mechanical applications, it wasn't clear how strong this effect would be and whether or not the Young's fringes exist for turbulent flows. It has been shown by Wernekinck *et al.* [63,144,250,251] that quantitative measurements in turbulent density fields can be performed by means of speckle photography. Such photography in double exposure mode requires the use of a very short light source, e.g. a Q-switched ruby laser. One obtains then, for one particular instant of time, *the instantaneous* distribution of deflection angles, $\epsilon(x, y)$, in whole field of view.

Using the advantage of speckle photography for quantitative measurements of light deflection angles, Erbeck and Merzkirch performed a statistical treatment of the turbulent density field of the flow behind the turbulence grid [340,361,362].

As a first step, one may calculate the correlation function of the deflection angle $\epsilon(x, y)$, and if isotropic turbulence can be assumed it is possible to determine the correlation function of the 3D, turbulent density field $\rho(x, y, z)$ and other characteristic properties of the turbulent field which are related to this function [363].

Theoretical description. Let us use the coordinate system (x, y, z) together with the system (p, q, s) where the direction of the s-axis coincides with that of the z-axis and the axes (p, q) might be rotated with respect to the (x, y) axes by the angle γ, see Fig. 9.3. The dimensions of (x, y, z) are in units of length and the dimensions of (p, q, s) are in pixel numbers. If the total number of pixels along these axes are N, M, and Q respectively, the correlation function for the density field in isotropic turbulence can be calculated using the formula

$$R_\rho(\tau) = \frac{1}{(N-\tau)^3} \sum_{p=1}^{N-\tau} \sum_{q=1}^{N-\tau} \sum_{s=1}^{N-\tau} \rho(p, q, s) \cdot \rho(p+\tau, q+\tau, s+\tau) \qquad (9.29)$$

where $Q = M = N$ and $\tau = \sqrt{\xi^2 + \eta^2 + \zeta^2}$.

For non-isotropic turbulence there are three correlation functions along the coordinate axes p, q, and s respectively:

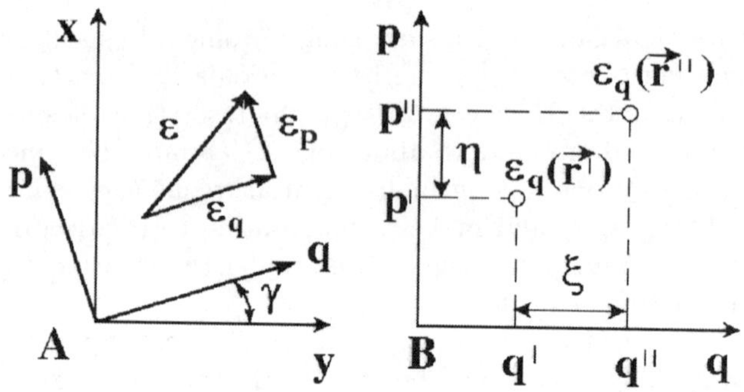

Fig. 9.3. Coordinate system for correlation function evaluation

$$R_\rho(\xi) = \frac{1}{(N-\xi)} \sum_{p=1}^{N-\xi} \rho(p,q,s) \cdot \rho(p+\xi,q,s) \tag{9.30}$$

$$R_\rho(\eta) = \frac{1}{(M-\eta)} \sum_{q=1}^{M-\eta} \rho(p,q,s) \cdot \rho(p,q+\eta,s) \tag{9.31}$$

$$R_\rho(\zeta) = \frac{1}{(Q-\zeta)} \sum_{s=1}^{Q-\zeta} \rho(p,q,s) \cdot \rho(p,q,s+\zeta). \tag{9.32}$$

Since the point-by-point reconstruction technique provides the speckle displacement $\Delta(x,y)$ by magnitude and direction, such an evaluation yields from the specklegram the two deflection angles $\epsilon_x(x,y)$ and $\epsilon_y(x,y)$ or $\epsilon_p(p,q)$ and $\epsilon_q(p,q)$ at the same time, in contrast to the Toepler schlieren system and shearing interference methods, all of which can measure, in a single experiment, only one component of the deflection angle [363]. Denoting by $\boldsymbol{\epsilon}(x,y)$ the light deflection field in the recording plane, orthogonal to the direction of the parallel light beam propagation, the vector value of deflection angle can be composed from two components, ϵ_p and ϵ_q :

$$\boldsymbol{\epsilon} = \mathbf{i} \cdot \epsilon_p + \mathbf{j} \cdot \epsilon_q. \tag{9.33}$$

The correlation function for the deflection angle field according to its definition is

$$R_\epsilon(p,q,\xi,\eta) = < \boldsymbol{\epsilon}(p,q) \cdot \boldsymbol{\epsilon}(p+\xi,q+\eta) > . \tag{9.34}$$

Here the symbol $< \ldots >$ denotes spatial averaging. For locally isotropic turbulence in principle it is possible to determine the spatial dependence of the correlation function, $R(p,q)$, but in practice it is necessary to perform spatial averaging of experimental data and due to a finite number of measurement points from one specklegram usually it is possible to reconstruct only a limited number or even one correlation function $R(\xi, \eta)$.

This function can also be composed from two parts, $R_{\epsilon_p}(\xi, \eta)$ and $R_{\epsilon_q}(\xi, \eta)$, where

$$R_{\epsilon_p}(\xi, \eta) = < \epsilon_p(p, q) \cdot \epsilon_p(p + \xi, q + \eta) > \tag{9.35}$$

and

$$R_{\epsilon_q}(\xi, \eta) = < \epsilon_q(p, q) \cdot \epsilon_q(p + \xi, q + \eta) > . \tag{9.36}$$

In turn the correlation functions $R_{\epsilon_q}(\xi, \eta)$ and $R_{\epsilon_p}(\xi, \eta)$ can be calculated in directions parallel (denoted by the symbol $\|$) or perpendicular (denoted by the symbol $\perp$) to the projections of the deflection angles

$$R_{\epsilon_q \perp}(\xi = \tau, \eta = 0) = R_{\epsilon_q \perp}(\tau); \qquad R_{\epsilon_q \|}(\xi = 0, \eta = \tau) = R_{\epsilon_q \|}(\tau) \tag{9.37}$$

and

$$R_{\epsilon_p \perp}(\xi = 0, \eta = \tau) = R_{\epsilon_p \perp}(\tau); \qquad R_{\epsilon_p \|}(\xi = r, \eta = 0) = R_{\epsilon_p \|}(\tau). \tag{9.38}$$

This means, for instance, that the parallel correlation function for ϵ_q is calculated along the q axis, and the perpendicular correlation function for the same function ϵ_q is calculated along the p-axis.

For a finite number of measured points (p_i, q_i) with $1 < p_i < m, 1 < q_i < n$ this correlation function can be approximated by the series

$$R_{\epsilon q \perp}(\tau) = \frac{1}{m - \tau} \sum_{i=1}^{m-\tau} \epsilon_q(i, q) \cdot \epsilon_q(i + \tau, q) \tag{9.39}$$

and

$$R_{\epsilon q \|}(\tau) = \frac{1}{m - \tau} \sum_{j=1}^{n-\tau} \epsilon_q(p, j) \cdot \epsilon_q(p, j + \tau). \tag{9.40}$$

We will see below that these correlation functions are not equivalent even for homogeneous turbulence.

Firstly let us show the relationship between the 3D density correlation function and the correlation functions for the 2D deflection angles field.

The relation between the deflection angle and the fluid density in the chosen coordinate system is

$$\epsilon_q(p, q) = K \int_0^L \frac{\partial \rho(p, q, s)}{\partial q} \, \mathrm{d}s \tag{9.41}$$

with L being the width of the test section in the z-direction. Therefore

$$R_{\epsilon_q}(\xi,\eta) =< K \int_0^L \frac{\partial\rho(p^{'},q^{'},s)}{\partial q}\,\mathrm{d}s \cdot K \int_0^L \frac{\partial\rho(p^{''},q^{''},s)}{\partial q}\,\mathrm{d}s > \tag{9.42}$$

where $p^{''} = p^{'} + \xi$ and $q^{''} = q^{'} + \eta$, and the vectors $\mathbf{r}^{'}$ and $\mathbf{r}^{''}$ are

$$\mathbf{r}^{'} = \begin{bmatrix} s^{'} \\ p^{'} \\ q^{'} \end{bmatrix} \tag{9.43}$$

and

$$\mathbf{r}^{''} = \begin{bmatrix} s^{''} \\ p^{''} \\ q^{''} \end{bmatrix} = \begin{bmatrix} s^{'} \\ p^{'} + \xi \\ q^{'} + \eta \end{bmatrix} = \begin{bmatrix} s^{'} + (s^{''} - s^{'}) \\ p^{'} + \xi \\ q^{'} + \eta \end{bmatrix}. \tag{9.44}$$

Using these relations, Erbeck and Merzkirch [340,361] obtained a connection between the density and the deflection angle correlations

$$R_{\epsilon q}(\xi,\eta) = -K^2 \int_0^L \int_0^L \frac{\partial^2}{\partial\eta^2} R_\rho(\xi,\eta,\zeta)\,\mathrm{d}z'\,\mathrm{d}z''. \tag{9.45}$$

Note that the correlation function of the density field depends on three spatial coordinates, namely ξ,η and ζ, where $\zeta = z^{''} - z^{'}$. Integration along the z-axis gives

$$R_{\epsilon q}(\xi,\eta) = -LK^2 \int_0^L \frac{\partial^2}{\partial\eta^2} R_\rho(\xi,\eta,\zeta)\,\mathrm{d}\zeta. \tag{9.46}$$

For isotropic turbulence the correlation functions depend on one parameter, τ or r, respectively:

$$\tau = \sqrt{\eta^2 + \xi^2}; \quad r = \sqrt{\tau^2 + \zeta^2} \tag{9.47}$$

and relation (9.46) can be rewritten in the form

$$R_{\epsilon q}(\tau) = -2K^2 L \int_0^L \frac{\partial^2}{\partial\eta^2} R_\rho(\sqrt{\tau^2 + \zeta^2})\,\mathrm{d}\zeta. \tag{9.48}$$

For the case of isotropic turbulence, this equation can be inverted with respect to correlation functions denoted by the symbols $\|$ and $\perp$:

$$R_\rho(r) = \frac{1}{\pi L K^2} \int_r^\infty \frac{1}{\sqrt{\tau^2 - r^2}} \{ \int_0^\tau R_{\epsilon q \|}(\tau^*)\, \mathrm{d}\tau^* \}\, \mathrm{d}\tau \tag{9.49}$$

and

$$R_\rho(r) = \frac{\tau}{\pi L K^2} \int_r^\infty \frac{1}{\sqrt{\tau^2 - r^2}} R_{\epsilon q \perp}(\tau)\, \mathrm{d}\tau \tag{9.50}$$

Using these correlation functions, the micro- and macro-scales of the turbulence may be evaluated, see Section 9.1:

$$L_\rho = \int_0^\infty \frac{R_\rho(r)\, \mathrm{d}r}{R_\rho(0)} \tag{9.51}$$

and

$$\lambda_\rho = -2 \frac{R_\rho(0)}{R''_\rho(0)} \tag{9.52}$$

with

$$R''_\rho(0) = \frac{dR_\rho}{dr}|_{r=0}.$$

Let us now calculate the deflection correlation functions or the density correlation function, expressed, e.g. as a Gaussian:

$$R_\rho(r) = R_\rho(0) \cdot \exp[-\left(\frac{r}{\lambda_\rho}\right)^2]. \tag{9.53}$$

Then the deflection angle correlations are

$$\begin{aligned} R_{\epsilon \|}(\tau) &= R_\rho(0) K^2 L \tfrac{\sqrt{\pi}}{\lambda_\rho} \\ & \times \left\{ 2 \exp[-\left(\tfrac{\tau}{\lambda_\rho}\right)^2] - 4 \left(\tfrac{\tau}{\lambda_\rho}\right)^2 \exp[-\left(\tfrac{\tau}{\lambda_\rho}\right)^2] \right\} \end{aligned} \tag{9.54}$$

and

$$R_{\epsilon \perp}(\tau) = R_\rho(0) 2 K^2 L \frac{\sqrt{\pi}}{\lambda_\rho} \exp[-\left(\frac{\tau}{\lambda_\rho}\right)^2]. \tag{9.55}$$

Thus it is clearly seen that even for homogeneous turbulence, the functions $R_{\epsilon \|}(\tau)$ and $R_{\epsilon \perp}(\tau)$ are different. It should be noted that inversion of this functions to the original one, $R_\rho(r)$, is an ill-posed mathematical problem and care must be taken in performing such calculations. As we can see, the inversion integrals are Abel type integrals, therefore the great experience of solving this equation may be used; see the next chapter.

Numerical simulation. Using the results of numerical ray tracing through the turbulence, see Section 9.1.2, the 3D density correlation function here was reconstructed with the help of the Erbeck–Merzkirch inverse integral transformation (9.49) using about 200 readings at each specklegram line. The results were compared with the 'initial' density correlation function, evaluated using the turbulent density distribution (see Fig. 9.4). The analysis shows that the quality of this reconstruction for small-scale vorticity is rather good with grids of about 200^2. The discrepancy between the initial correlation function constructed using the results of direct numerical simulation and the function constructed using the deflection angle map increases with increasing vorticity scale.

These calculation were made for non-isotropic turbulence with a value of anisotropy of about 15%. In spite of this it is possible to see that the precision of turbulence microscale determination using the integral transform for the isotropic turbulence is rather high and the error of the determination is less than 5%. It should be noted that this precision strongly depends on the experimental data density and goes down rapidly with experimental grids less than about 100^2.

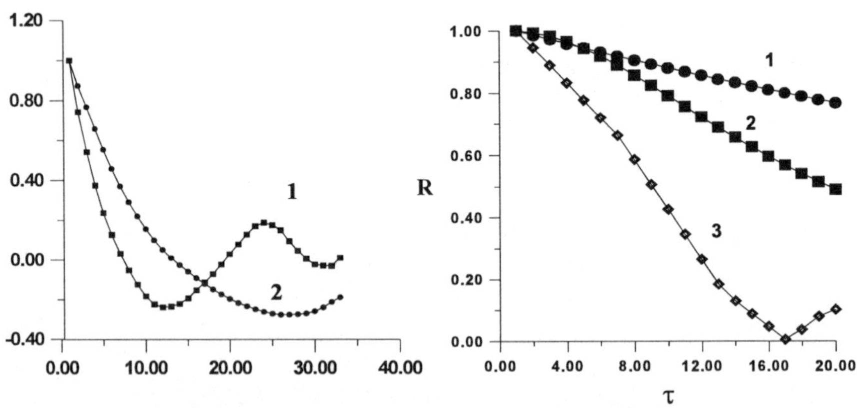

Fig. 9.4. Left, correlation functions of deflection angles. 1, orthogonal and 2, parallel correlation functions. Right, correlation functions reconstructed from computer-simulated specklegram. 1. The density correlation function, reconstructed using the 2D deflection angle map, obtained by specklegram treatment. 2. The 'initial' density correlation function, obtained by using the results of direct numerical simulation of turbulence. 3. The deflection angle correlation function.

Experimental verification. The first experimental implementation of the technique described above was realized in publications [214,340,361].

The experiments have been performed in a small, low-speed wind tunnel with a quadratic test section of $457 \times 457\,\text{mm}^2$. Nearly isotropic turbulence was generated by a grid of mesh size $M = 41mm$. One mesh size downstream of the turbulence grid a system of electrically heated wires was placed. The wire spacing was equal to M. This heating of the air flow results in mean values of the temperature fluctuation of the order of 1 K. Thus, geometry, heating system, and flow conditions closely match the conditions of a number of reported experiments in which the temperature fluctuations were measured with a cold-wire probe.

These data as well as the comparison of turbulence microscale determinations show that the speckle measurements are in good agreement with respective cold-wire measurements. It should be noted that the speckle measurement was performed *instantaneously* during the Q-switched ruby laser exposure (20 ns), whereas the measurement with the help of cold-wire anemometry needs to use a slow scanning system.

This technique was used later for the evaluation of interactions of the turbulence with the combustion front and with the shock wave, see Merzkirch and Oberrste–Lehn [342,364], Merzkirch and Keller [341]. The energy spectrum of turbulent density fluctuations was calculated using Fourier analysis of the obtained correlation functions.

In the next section this analysis will be applied to the study of interactions of the turbulence field with shocks.

9.3 Turbulence structure visualization in flows with strong anisotropy

The speckle photography method can be applied to studies of both homogeneous isotropic turbulence and turbulent flows with strong anisotropy. In the latter case, however, the construction of the density correlation function using single-projection measurements and integral transformations (9.49)–(9.50) could give rise to inaccuracies in the reconstruction. A better approach in this case is multi-projection measurements and reconstruction of the local turbulence parameters using computerized tomography which is the subject of the next chapter. Nevertheless, even in the case of strong anisotropy, it is possible to obtain useful information about turbulence using single-projection data. As shown by Vitkin *et al.* [365] even in this case direct analysis of the deflection angle correlation function obtained by using double-exposure speckle photography provides quantitative turbulence structure visualization.

A good example of turbulence with strong anisotropy is turbulent flow after interaction with a planar shock wave. There are both theoretical [366-368] and experimental [112] indications that such flow is far from being isotropic. The results of quantitative flow visualization confirm and demonstrate the high potential of speckle photography for turbulence structure studies.

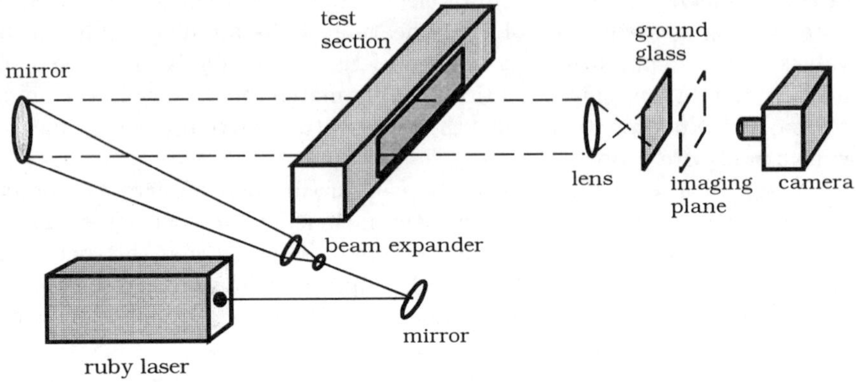

Fig. 9.5. Optical set-up for taking speckle photographs in a shock tube experiments (from Vitkin *et al.* [365])

9.3.1 Speckle photography of shock tube generated flow

The experiments were performed with unsteady air flow in a shock tube at Essen University, having a quadratic cross-section of $100 \times 100\,\mathrm{mm}^2$. The shock propagates into the tube of quadratic cross-section where it reflects at the closed end. Integrated in the mechanism used for destroying the diaphragm is a turbulence grid, so that the air expanding from the compression tank must pass through the grid. Thereby, a turbulent air flow with density fluctuations is generated in the shock tube, and the front of the turbulent regime coincides with the contact front that moves with the local air velocity and separates the air which was originally in the low pressure side from that in the compression tank. At a distance of 0.2 m downstream of the diaphragm/turbulence grid is the optical test section with transparent vertical side walls made of high-quality optical glass mounted flush with the plane inner side walls of the shock tube. The size of the glass window is $200 \times 100\,\mathrm{mm}^2$. The expanded parallel laser beam used in the speckle photographic set-up (see below) is directed through the windows of the optical test section, normal to the tube axis.

The optical set-up shown in Fig. 9.5 is that described by Wernekinck and Merzkirch [145]. The beam from a pulsed ruby laser is expanded and collimated to form a parallel beam of 10 cm diameter. After having passed horizontally through the optical test section of the shock tube, the laser light illuminates a disk of ground glass, the speckle generating element in the set-up. The plane in the speckle field, normal to the optical axis and at a distance of about 10 mm from the ground glass, is imaged on to the photographic plate where the speckle patterns are recorded. We image a field of view of about 10 cm in diameter with an imaging ratio of 1:1 on to the photographic plate.

The first exposure is taken without flow and serves as the reference speckle pattern. The second exposure with flow, recorded on the same plate, is taken at a desired instant of time as defined by the instantaneous position of the moving shock wave. Due to variations of the fluid density occurring in the flow the light rays in the second exposure are deflected with respect to the direction of light propagation in the reference exposure, thus resulting in a (locally inhomogeneous) displacement of the second speckle pattern in comparison to the reference pattern. The double-exposed speckle photograph is developed and then interrogated with a thin He-Ne laser beam in order to determine the local speckle displacement $\Delta(x, y)$, respectively the local light deflection angle $\epsilon(x, y)$, via the method of Young's fringes. This evaluation is performed with an automated system that, with the presently used hard- and software, can evaluate 2500 data points per hour. The major part of the evaluation time is needed for moving the interrogating laser beam mechanically, i.e. by stepping motors, from one interrogation point to the next. In most cases we evaluate a specklegram on a grid of 100 $\times$100 data points, so that approximately 4 hours are needed for the analysis of one specklegram (double-exposure). The interrogation is done in steps of 0.2 mm which corresponds to the diffraction limit of spatial resolution in our system, given by $\sqrt{\lambda L}$, with λ being the laser wave length and L the depth of the optical test section in the direction of light propagation (z-axis).

9.3.2 Results of the diagnostics

With the pulse length of the illuminating ruby laser being approximately 50 ns, we freeze in the specklegram the instantaneous distribution of the deflection angles $\epsilon(x, y)$ as caused by the turbulent density field. A specklegram that includes the shock wave propagating through the turbulent field is shown in Fig. 9.6.

The contrast of this specklegram was enhanced by applying a digital filter, and it resembles the patterns of turbulent compressible flows as visualized by shadowgraphy. Due to its interaction with the density fluctuations the shock is no longer perfectly plane. An instantaneous distribution of the deflection angles $\epsilon(x, y)$ is presented in Fig. 9.7 in the form of isolines. The two fields of view shown in Fig. 9.7 are the distributions before (left) and after (right) passage of the shock. The distance from the instantaneous position of the shock wave is indicated. The visible patterns can be interpreted as turbulent structures, although one must be aware of the integrating effect of the optical line-of-sight method. Evidently, the structures are more extended in the y-direction than in the x-direction (direction of flow), which can be taken as an indication of the anisotropy of the turbulent field. This anisotropy has also been verified with single-exposure speckle-photographic experiments [112,114]. A further observation is that the amount of light deflection increases after the passage of the shock (right side in Fig. 9.7), which means that the turbulence intensity increases.

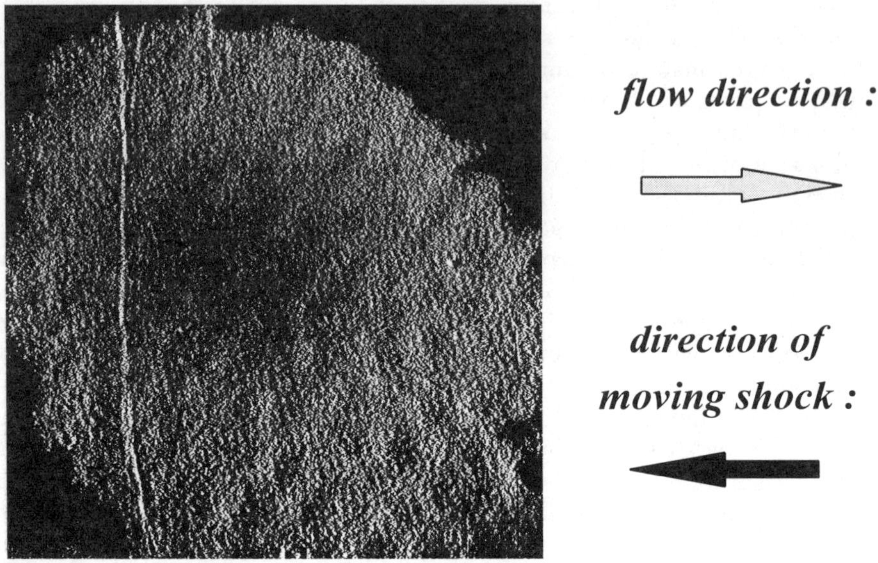

Fig. 9.6. Speckle photograph showing a wave propagating normally through the turbulent gas flow with fluctuating density (from Vitkin *et al.* [365])

But in the field of view downstream of the shock wave included in Fig. 9.7 the turbulence is certainly not in an equilibrium state. It is therefore of interest to study the turbulent field at larger distances from the shock. Such a study was performed by Vitkin *et al.* [365]. In this work the specklegrams were taken (a) 0.26 ms prior to the arrival of the shock (left), (b) 0.5 ms after the passage of the shock (middle) and (c) 2.2 ms after the passage of the shock (right). The centres of the three sections shown are at a distance from the moving shock of approximately 10 cm (left), 20 cm (middle) and 88 cm (right). There appears to be a small increase of the size of the visible turbulent structures from position (a) to (b). But at the larger distance (c) from the shock wave the structures are definitely smaller than they are ahead of the shock at position (a), and the turbulence level is also lower at (c) in comparison to (a). These observations can be quantified by determining the spatial correlation function R_ϵ of the deflection angles in the (x, y) plane according to Eq.(9.34). The non-dimensional spatial correlation functions for the two cases shown in Fig. 9.7 are presented in Fig. 9.8.

A comparison of the correlations in the x- and y-directions reveals the following results:

– The obvious elongation of the turbulent structures in the y-direction for cases (a) and (b) is expressed by the higher values of the correlation functions in this direction. This significant difference of the correlations in the

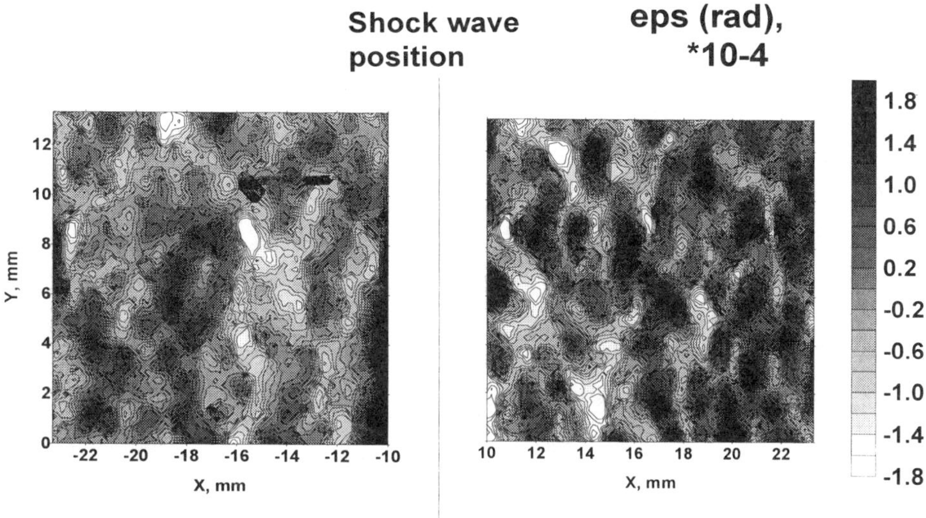

Fig. 9.7. Instantaneous distribution of the light deflection angles $\epsilon(x, y)$ at two positions downstream (left) and upstream (right) of the moving shock. The shock propagates in the negative x-direction. X-values refer to the distance from the instantaneous shock position. Deflection angles ϵ are measured in radians

x- and y-directions confirms the high degree of anisotropy of the turbulent scalar field at these positions.

- At a larger distance behind the shock (position c) the turbulence is almost isotropic. The more rapid decrease of the correlation in the y-direction, as compared to cases (a) and (b), is an indication of the decrease in size of the turbulent structures.

The obtained results show that the spatial structure of the turbulent scalar (density) field in this compressible flow can be visualized quantitatively with the applied optical specklegram technique. The patterns visible in the distributions of the deflection angle isolines are interpreted as large-scale turbulent structures. The turbulence in the flow ahead of the shock is anisotropic, and it remains so for the first short period after the passage of the shock wave, with a slight increase in size of the structures. It is evident that different states of turbulence develop downstream of the shock, depending on the out-of-equilibrium degree as caused by the shock compression. The tendency towards an isotropic state at some distance from the shock is understandable, because the flow velocity behind the reflected shock wave is nominally (under pure gas-dynamic aspects) zero. In our interpretation of the visible results, however, we have to pay attention to the fact that all optical signals are integrated along the path of the light through the test section. A direct comparison with results of numerical simulations, e.g. of Lee *et al.* [366], is therefore not possible. Such a comparison would require a conversion

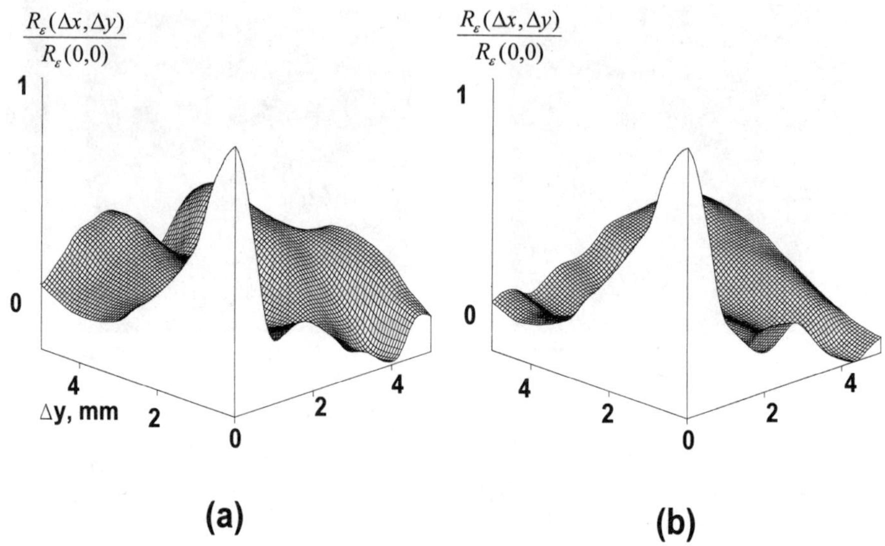

Fig. 9.8. Two-dimensional correlation functions of the light deflection angles for the two cases shown in Fig. 9.7: (a) 10 cm ahead of the shock (or 0.26 ms before the arrival of the shock), and (b) 20 cm behind the shock ($\sim$ 0.5 ms after the arrival of the shock)

of the 2D optical data into 3D quantities describing the turbulence characteristics of the density field. A possible approach to such conversion is the subject of the next chapter.

9.4 Single exposure technique

The recording of a moving speckle pattern over a prolonged time has been used in speckle diagnostics from the very beginning of its development. Thus, Tiziani describes *the time-averaged speckle photography* for analysing vibration systems [101,102].

Fercher and Briers developed *the single exposure speckle photography* method for quantitative *random* velocity determination by mapping the variation of speckle contrast [104]. In particular, this technique was applied to the visualization of blood flow in the human retina. Later, an optical filtering technique was used to enhance speckle contrast variations in single-exposure speckle photography [107], and a quasi-real time digital version of the method, based on CCD camera recording, was implemented [108].

Barakat *et al.* [110] have used single exposure speckle photography for regular velocity determination in very slow flow in a circular pipe. The diffraction halo analysis was used here for the flowfield reconstruction. A similar approach was applied to angular velocity determination of a rotating disk [109].

In this section *a line-of-sight* single exposure speckle photography technique *in a frame mode* will be described allows one to measure the degree of anisotropy and the level of turbulence in a flow with fluctuating fluid density [114]. The extension of this technique to *a streak mode* allows one to determine the temporal evolution of the level of turbulence along selected lines in the flow [117].

9.4.1 Experimental set-up

The experimental installation for SPESP is much the same as that for DESP (see, e.g. Fig. 9.9).

A pulsed ruby laser without intensity modulation (pulse duration about 0.5 ms) and CW lasers with mechanical (camera) shutters (pulse duration exceeding 0.5 ms) are used as radiation sources. A collimated laser beam passes via a test object and is directed to the scattering plate (ground glass) which is the speckle generator. A speckle pattern is formed on the photographic plate plane by a camera lens and recorded over a time that is long compared to the time of the double exposure speckle photography technique.

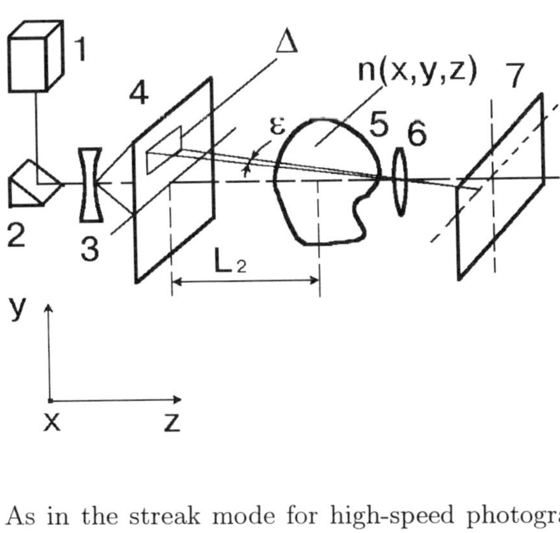

Fig. 9.9. Experimental set-ups for SPESP in frame mode. 1, laser with relatively long pulse, 2, reflecting mirror, 3, beam expander, 4, ground glass (speckle generator), 5, turbulent medium under study, 6, imaging lens, 7, photographic plate

As in the streak mode for high-speed photography, a slit is installed to record the time history of a single exposure speckle pattern along the selected line in the flow studied. Individual speckles in such modes will form on the moving film streaks with different thickness, similar to interferometric fringes. The thickness of these fringes is typically about 1–10 μm in size, the same as the typical speckle size for the SPESP scheme. This thickness at a chosen film position will be determined by the intensity of density (temperature) fluctuations at a chosen instant of time.

In frame mode the SPESP can give additional information to DESP about turbulence parameters of the flow under study. In this paper the whole field of the intensity of temperature fluctuations in a turbulent flame was measured by SPESP. It is shown that in some flow regions these fluctuations are different in different directions in the flow. So, the value and direction of turbulence anisotropy was quantitatively measured and the 2D maps of the turbulence anisotropy were also presented.

Both in frame and in streak mode, when the fluctuating speckle pattern is recorded photographically over a time that is long compared to the characteristic gas-dynamic time and the laser beam forming the speckle pattern passes through a turbulent flow, the images of the speckles are smeared. This results in a relatively large speckle size. The speckle enlargement is correlated to the intensity of temperature fluctuations in the turbulent flow under study and can be reconstructed through analysis of the specklegram diffraction halo, whose light distribution is proportional to the Fourier transform of the optical density distribution within the treated specklegram point.

The optical system for single exposure specklegram evaluation is also essentially the same as that for double exposure specklegram treatment. The interrogation of the specklegram with a thin laser beam results in the formation of a diffraction halo, whose light distribution is proportional to the Fourier transform of the optical density distribution within the treated specklegram point. While in DESP the turbulent speckle motion is frozen because of very short exposure time, the halo distribution is modulated by Young's interferometric fringes whose spacing is correlated to the value of the speckle shift at the treated point of the specklegram during the time interval between two exposures. The halo distribution for the SPESP is not modulated by Young's fringes (see Fig. 9.10) and has a circular or elliptical form. Usually only the first order of diffraction can be observed.

The halo cross-section is circular if turbulent fluctuations in a plane normal to the direction of light propagation are evenly distributed, as in isotropic turbulence. With anisotropic turbulence, the speckle becomes elongated in the preferential direction of turbulence, and the halo cross-section is elliptical with the longer axis being normal to the direction of the elongated speckles.

9.4.2 Diffraction halo analysis

The averaged speckles and observed halo are both of circular cross-section if the turbulent fluctuations are uniformly distributed in a plane normal to

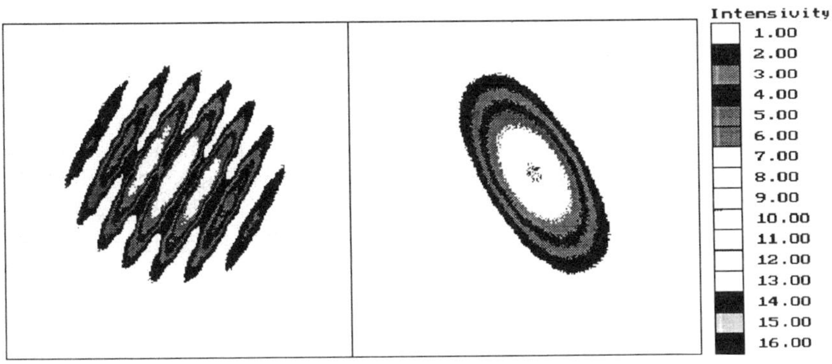

Fig. 9.10. The laser light intensity distribution in the diffraction halo region during specklegram processing; left DESP, right SPESP

the direction of light propagation, as given in isotropic turbulence. The halo diameter D, determined, e.g., at half of the intensity maximum, is inversely proportional to *the averaged* speckle diameter $< d >$. With anisotropic turbulence, the speckles become *on the average* elongated in the preferential direction of the turbulent fluctuations, and the halo then has an elliptical cross-section, with the longer axis of length D being normal to the direction of the speckle elongation (Fomin *et al.* [111]). The *averaged* deviation of the speckle from circular form is expressed by the quantity $< \Delta d >$ (Fig. 9.11).

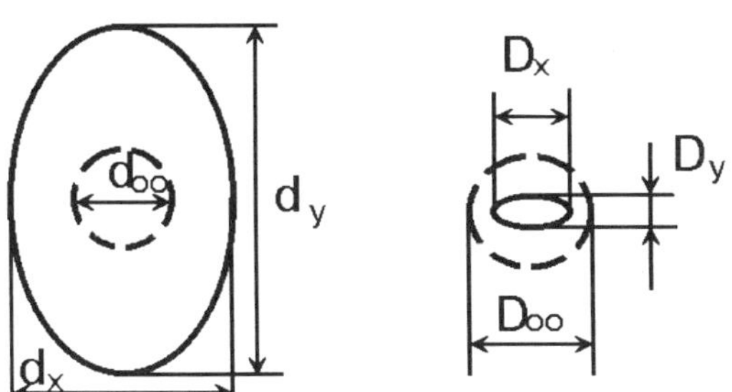

Fig. 9.11. A model of the averaged speckle form and diffraction halo cross-section for undisturbed specklegram regions (doted) and specklegram regions dissturbed by flow (solid)

If several elongated speckles, all of them being deformed into the same direction, are present in an interrogation area, the extension D of the resulting halo is given by

$$\sqrt{\frac{< (\Delta d)^2 >}{< (d_\infty)^2 >}} = C_1 \cdot (\frac{D_\infty}{D} - 1), \tag{9.56}$$

where C_1 is a constant, $< d_\infty >$ is the averaged diameter of speckles in the fluid without flow, and D_∞ is the halo diameter for the speckles of size $< d_{\infty} >$. The average $< \ldots >$ is taken over all speckles inside the interrogation area.

With respect to the 'undisturbed' speckle of diameter d_∞ the speckles might be deformed by the turbulent density fluctuations in two orthogonal directions x and y, but by different amounts, Δd_x and Δd_y (Fig. 9.11).

The lengths of the two axes, D_x and D_y, of the resulting elliptic halo cross-section are given by

$$\sqrt{\frac{< (\Delta d_x)^2 >}{< (d_\infty)^2 >}} = C_1 \cdot (\frac{D_\infty}{D_y} - 1), \tag{9.57}$$

$$\sqrt{\frac{< (\Delta d_y)^2 >}{< (d_\infty)^2 >}} = C_1 \cdot (\frac{D_\infty}{D_x} - 1). \tag{9.58}$$

The quantity A describing the deviation of the halo cross-section from circular form can be defined as

$$A = \sqrt{\frac{< (\Delta d_x)^2 >}{< (\Delta d_y)^2 >}} = (\frac{D_\infty}{D_x} - 1)/(\frac{D_\infty}{D_y} - 1) \simeq \frac{D_y}{D_x}, \tag{9.59}$$

where $(D_x - D_y) << D_\infty$ has been assumed in the last relationship.

The averaged speckle elongation, expressed by Δd, is related to the average of the square of the light deflection angle fluctuations, $\Delta\Theta$, by

$$< \Delta d^2 >= L_1^2 \cdot < \Delta\Theta^2 > . \tag{9.60}$$

In the system described here L_1 is the 'defocusing' distance between the ground glass and the plane imaged on to the photographic plate.

As shown by Weiner [369], the mean square of the temporal fluctuations of the deflection angle is related to the mean square of the refractive index (n) variations by

$$\overline{\Delta\Theta^2} = \frac{2}{3} \cdot \frac{L}{L_C} \cdot \overline{\Delta n^2}, \tag{9.61}$$

where L is the length of the optical path through the flow under study, and L_C is a correlation length of the turbulent density field, e.g. the turbulent integral scale. The upper bar refers to time averaging.

When using the ergodicity hypothesis [303], the statistical mean of the temporal variations is equal to the statistical average of the spatial variations:

$$\overline{\Delta n^2(t)} =< \Delta n^2(r) >, \tag{9.62}$$

and

$$\Delta n = n - \overline{n} = n - < n > . \tag{9.63}$$

For isobaric flows the variation of the refraction index n is related to the variation of temperature by

$$< \Delta n(r)^2 >= C_2 \cdot < \Delta T(r)^2 > \tag{9.64}$$

with C_2 being constant. The mean square value of speckle elongation in the Ith direction is then given by

$$< \Delta d_{\mathrm{i}}^2 >= C_3 \cdot < \Delta T_{\mathrm{i}}^2 >, \tag{9.65}$$

where C_3 is constant. Therefore, to a first approximation, the measured value of $(D_\infty / D - 1)$ is proportional to the intensity of the temperature fluctuations:

$$D_\infty / D - 1 \sim \sqrt{\frac{< \Delta T^2 >}{T_\infty^2}}, \tag{9.66}$$

and the quantity $A = D_x / D_y$ is proportional to the anisotropy of the turbulent temperature field:

$$A = D_x / D_y \sim \sqrt{\frac{< \Delta T_y^2 >}{< \Delta T_x^2 >}}. \tag{9.67}$$

9.4.3 Experimental verification

The cross-section of a halo projected on to a screen in the automated specklegram treatment installation described in Section 6.1.8 is viewed by a CCD camera whose recording is digitized and then evaluated for determining the quantities D (or D_x, D_y) and A. A digitized pattern of an elliptic halo and its intensity distribution are shown in Fig. 9.12.

Software for use on a PC was developed that allowed us to determine automatically the quantities D and A while the specklegram is interrogated with the He/Ne laser beam. RMS values of the temperature fluctuations and the anisotropy of the temperature field can be determined with the aid of Eqs. (9.66) and (9.67). The results of measurements were compared with hot wire anemometry data (see Fig. 9.13). The flow was Bashir and Uberoi's two-dimensional, non-isothermal turbulent jet [370]. A linear calibration curve was obtained in the range of rms temperature fluctuations up to 50 K. Speckle photography was able to record temperature fluctuations with a lower limit of approximately 1 K. Fig. 9.14 contains the quantitative rms temperature fluctuations, as well as the isolines of turbulence anisotropy, obtained for a turbulent flame from a single exposure speckle photograph.

Using these data, the correlation function of the fluctuating temperature field was also constructed [371]. In contrast to the correlation functions obtained using data from DESP, in the case of SPESP, no additional spatial

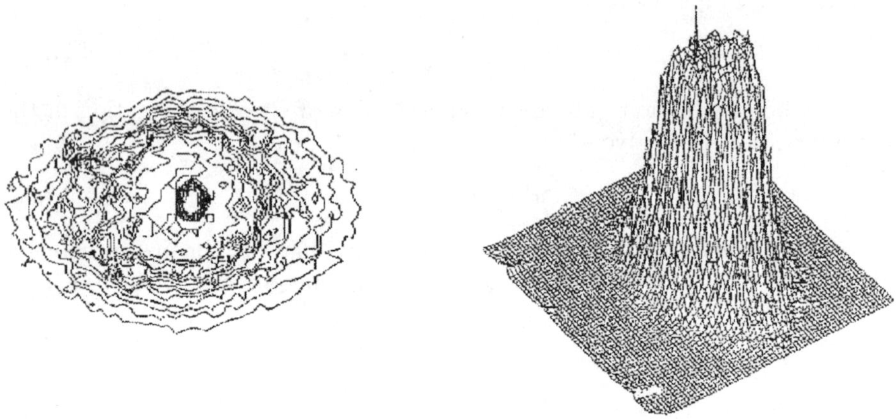

Fig. 9.12. Intensity distribution in the halo region for single exposure speckle photography (from Fomin *et al.* [112])

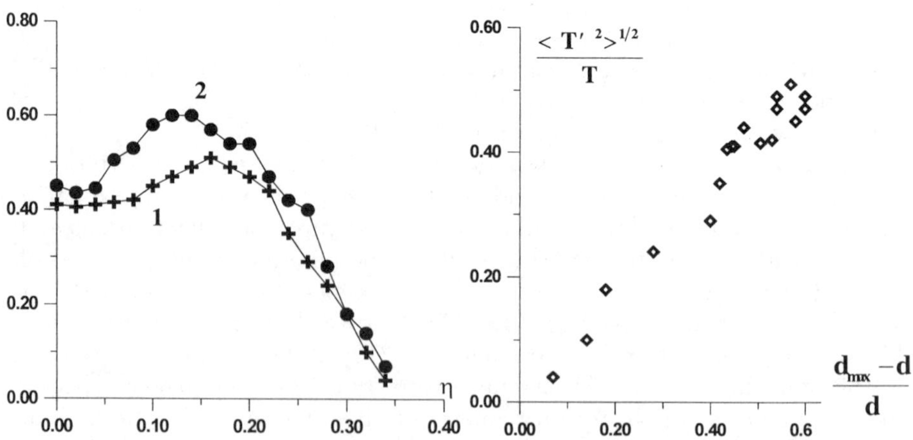

Fig. 9.13. Comparison of the variation of the halo diameter in speckle photography (1) with hot wire anemometry data about temperature fluctuations (2) in turbulent flow. The results of the calibration are shown on the right

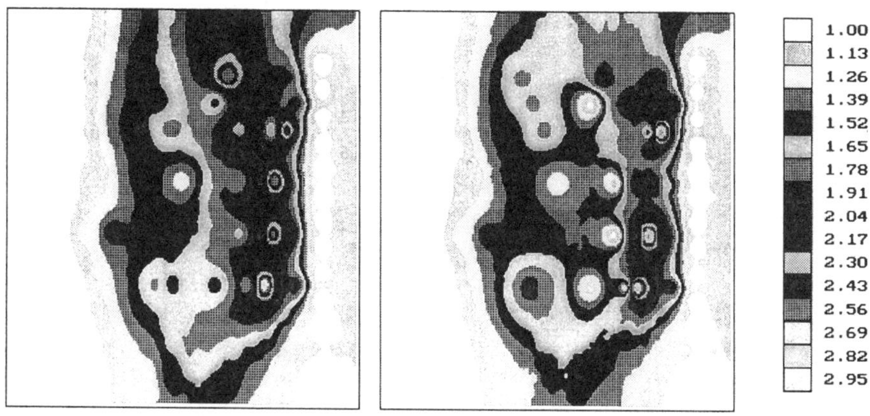

ture fluctuations (left) in the turbulent flame obtained by SP LSI

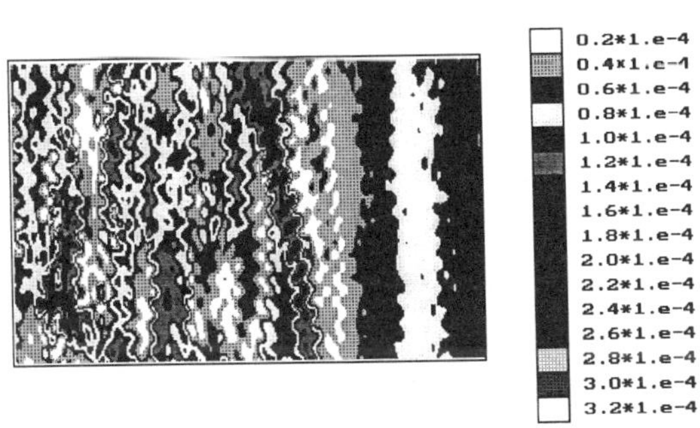

Fig. 9.15. The results of numerical filtering of a fragment of the specklegram corresponding to a region with strong anisotropy of turbulence

integration is needed. Therefore, using this approach the correlation function can be directly obtained with significantly better spatial resolution in the vicinity of each interrogation point.

Figure 9.15 contains the results of numerical filtering of a small fragment of a specklegram [372]. The flowfield is a connective zone of an open turbulent flame.

It is seen that for this instant at which the specklegram was recorded the turbulence to the right of the flow axis has a higher degree of anisotropy than that on the left side.

To summarize, single exposure speckle photography offers a means for quickly visualizing and determining the degree of anisotropy of a turbulent temperature or density field. Since anisotropy in linked to shear in a flow, the method is also a means for discriminating regions of different shear rate.

The SPESP technique can be used both in frame and in streak recording mode [373]. As in streak mode for high speed photography, a slit (see Fig. 5.3) was installed to record the time history of a single exposure speckle pattern along a selected line in the flow studied. Individual speckles in such modes will form on the moving film streaks with different thickness, similar to interferometric fringes. The thickness of these fringes is typically about 1–10 μm in size, the same as the typical speckle size for the SPESP scheme. This thickness at a chosen film position will be determined by the intensity of density (temperature) fluctuations at a chosen instant of time, see Fig. 9.16.

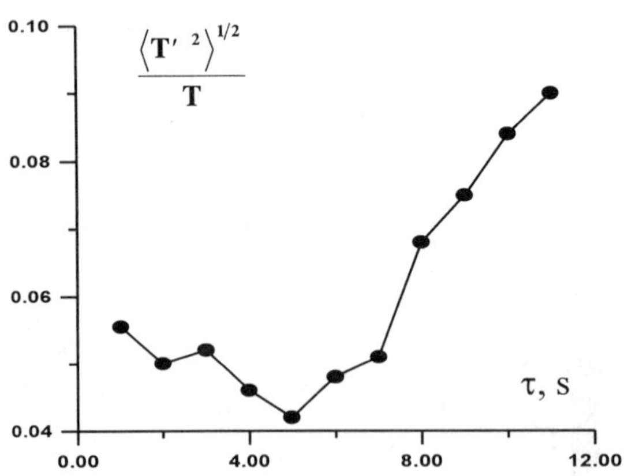

Fig. 9.16. Temporal evolution of temperature rms in a turbulent jet obtained by SPESP in streak mode

In frame mode the SPESP can give information additional to DESP about turbulence parameters of the flow under study. In [114] the whole field of the intensity of temperature fluctuations in a turbulent flame was measured by SPESP. It is shown that in some flow regions this fluctuation is different at different directions in the flow. So, the value and direction of turbulence anisotropy was quantitatively measured and the 2D maps of the turbulence anisotropy were also presented.

Both in frame and in streak mode, when the fluctuating speckle pattern is recorded photographically during times long compared to characteristic gas-dynamic time and the laser beam forming the speckle pattern passes through a turbulent flow, the images of the speckles are smeared. This results in a relatively large speckle size. The speckle enlargement is correlated to the intensity of temperature fluctuations in the turbulent flow and can be reconstructed through analysis of the specklegram diffraction halo, whose light distribution is proportional to the Fourier transform of the optical density distribution

As we can see, the single (prolonged) exposure speckle photography technique in the frame mode can both quantitatively reconstruct the two-dimensional field of turbulent temperature fluctuations and quantitatively visualize the value of turbulence anisotropy at the selected points of the flow as well as within the entire flowfield at the selected instances of time.

10. Speckle Tomography of Transparent Objects

Computerized tomography (CT) has become very well known due to advances in medical imaging. Tomography allows the distribution of physical parameters of a three-dimensional object to be determined from line-of-sight measurements. Since the early 1960s the tomographic approach has been applied to industrial imaging. A wide variety of problems including the nondestructive inspection of materials to locate defects, the determination of the distribution of dye in a cell, and *in vivo* measurements of metabolic activity in tissue, are examples of CT applications. Among other areas of applications there are radio astronomy, electron micrography, geology, oceanography, *etc.* The number of publications describing CT is enormous. Many books, review, and tutorial articles are devoted to various aspects of CT [374-378].

For fluid mechanical measurements CT was introduced as a technique that allows one to determine temperature, density, or concentration profiles in 3D flows by means of multi- projectional interferometric measurements [379-383].

A number of other techniques, based on a variety of physical phenomena, have been combined with CT for the study of 3D flows. Examples include the measurements of absorption [384-388], light emission [389], holographic interferomerty [390-392], and the speckle technique [146,393-397].

The first experimental demonstration of the tomography approach for asymmetric flow measurements was made by Matulka and Collins [381] who demonstrated the measurement of density in an air jet.

The information obtainable from a line-of-sight projection measurement consists of the averaged data along the path length of the particular ray s_{i}

$$\Phi_{\mathrm{s}_{\mathrm{i}}} = \int_{\mathrm{s}_{\mathrm{i}}} f(x, y, z)\, \mathrm{d}s_{\mathrm{i}}, \tag{10.1}$$

where $f(x, y, z)$ is the required function corresponding to the chosen optical scheme and $\Phi_{\mathrm{s}_{\mathrm{i}}}$ is the projection information for the particular ray s_{i}.

For instance, in the case of reference beam interferometry with finite fringe alignment, this is

$$\Phi_{S_{\mathrm{i}}} = \frac{\delta S(x, y)}{S}, \qquad f(x, y, z) = \frac{1}{\lambda}(n(x, y, z) - n_0); \tag{10.2}$$

in the case of absorption measurements it is

$$\Phi_{\mathrm{S_i}} = -\ln(\frac{I}{I_0}), \qquad f(x,y,z) = N_{\mathrm{i}} Q_{\nu i}, \tag{10.3}$$

where N_{i} is the number density of species i which absorbs at frequency ν, $Q_{\nu i}$ is the absorption cross-section for species i at frequency ν, and in the case of speckle photography it is a vector function:

$$\boldsymbol{\Phi}_{\mathrm{S_i}} = \epsilon_x \mathbf{i} + \epsilon_y \mathbf{j}, \qquad \mathbf{f}(x,y,z) = \frac{1}{n}(\frac{\partial}{\partial x}\mathbf{i} + \frac{\partial}{\partial y}\mathbf{j}) n(x,y,z). \tag{10.4}$$

Equation (10.1) may be considered to be a functional equation from which local values of the function $f(x, y, z)$ can be determined when the path length function Φ_{i} is known for a large collection of rays. Under the assumption of geometrical optics, this equation has to be solved with the eikonal equation describing the light rays traces.[1]

The general solution to the coupled non-linear set of equations (10.1) and the eikonal equation is not known. There have been many attempts to solve this equation set numerically, see e.g. [398]. By and large, the underlying approximation in the development of the reconstruction algorithms is that the probing radiation travels along a straight line, and the refraction effects are taken into account as corrections. If the bending of the probing rays due to refraction is not negligible, then Radon inversion is an unsatisfactory approximation and additional considerations are needed. There are computational schemes based on iteration algorithms [399], perturbation approach [400], or exact solutions of particular problems [401].

For the case of negligible refraction, it is possible to construct 3D flow distributions by combining the information obtained from a number of 'independent' 2D reconstructions. For example, given a vertical jet, it is possible to find the 3D distribution by reconstructing the distributions in a number of horizontal planes cutting through the jet. These planes may be spaced as closely as desired to give more or less spatial resolution in the vertical direction. These planar reconstructions may then be stacked to provide a 3D field [392]. So, we can consider below the problem of the reconstruction of some physical property in a plane cutting a 3D region without loss of generality, see Fig. 10.1.

Another approach, vertical cutting into sections, may be employed in some measurement schemes, like PIV, PLIF, RELIF, when direct recordings of the flow pattern in the transverse direction are possible (see Fig. 5.8). Here the local data distributions in the vertical flow cross-section are measured and no computerized reconstruction is needed. Such data is not disturbed by

[1] Strictly speaking, for speckle photography the Eq. (10.1) can de derived from the eikonal equation. Therefore, only the eikonal equation has to be solved to reconstruct the local parameters in the case of speckle photography; see Section 10.4.

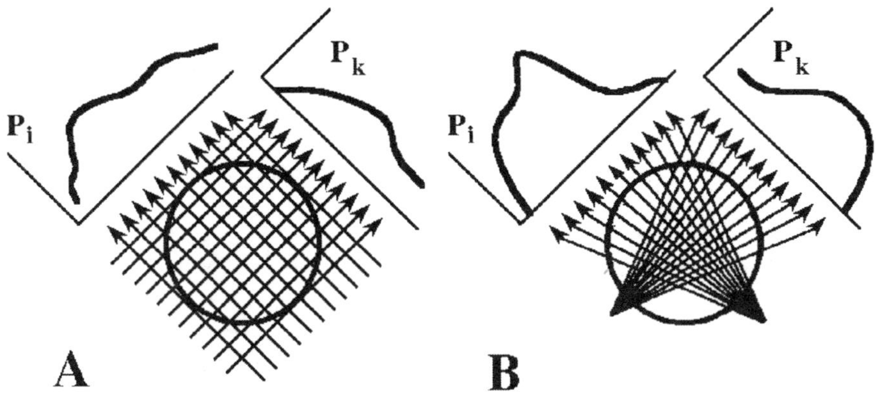

Fig. 10.1. Different possible scanning for tomographic measurements: A – parallel projection, B – fan-beam projection

integration along the optical path, and the whole 3D flow pattern can be constructed simply by combining the distributions in a number of sections [402].

For the case of relatively small refraction effects, it is convenient to introduce the 'probing direction', which may be characterized by a projection angle Θ. Then the projection, Φ_{S_i}, will be a function of two parameters, Θ and a coordinate p, describing the position of optical trace, and Eq. (10.1) may be rewritten

$$\Phi_\Theta(p) = \int\int f(x,y)\delta(x\cos\Theta + y\cos\Theta - p)\,\mathrm{d}x\mathrm{d}y \qquad (10.5)$$

where the integration is carried out along the line $x\cos\Theta + y\cos\Theta = p$, δ is the delta-function and $\Phi_\Theta(p)$ denotes the projection information in the Θ direction.

Using a polar coordinate system (r, ϕ) Eq. (10.5) transforms to

$$\Phi_\Theta(p) = \int_0^{2\pi}\int_0^{\infty} f(r,\phi)\delta[r\cos(\Theta-\phi) - p]r\,\mathrm{d}r\mathrm{d}\phi \qquad (10.6)$$

with the integration line defined by $r\cos(\Theta - \phi) = p$.

10.1 Abel and Radon inversions

Let us begin with a special case and assume that the 3D flow has rotational symmetry. Such flows, for example, are axisymmetric jets, flames, free convection along a vertical heated cylinder, or flow over an axisymmetric configuration [12]. Strictly speaking, the flow must be laminar, but the same analysis is often applied for turbulent flow too, see e.g. [391]. The applicability of the tomography approach for turbulent flows study will be discussed later.

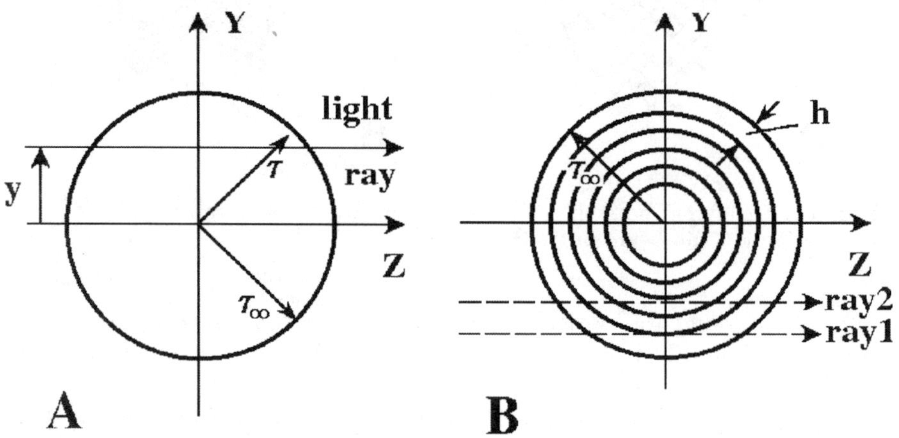

Fig. 10.2. Geometry for axisymmetric flow (A) and the ray path through annular zones (B)

We take the x-axis as the axis of symmetry, and the z-axis as the optical axis (Fig. 10.2). In this case for any fixed x

$$n(y,z) = n(r), \qquad r = \sqrt{y^2+z^2}, \qquad \mathrm{d}z = \frac{r\mathrm{d}r}{\sqrt{r^2-y^2}}. \tag{10.7}$$

Equation (10.1) is transformed into

$$\Phi(y) = \int_{r=y}^{r_\infty} f(r) \frac{\mathrm{d}(r^2)}{\sqrt{r^2-y^2}}. \tag{10.8}$$

This equation is referred to as the Wolter integral equation of the first type and is called an Abel integral. The equation can be rewritten in matrix form as

$$\Phi(y) = \mathcal{A}\{f(r)\}. \tag{10.9}$$

This equation has an analytical inversion

$$f(r) = \mathcal{A}^{-1}\{\Phi(y)\}, \tag{10.10}$$

where

$$\mathcal{A}^{-1}\{\Phi(y)\} = -\frac{1}{\pi}\frac{\mathrm{d}}{\mathrm{d}(r^2)}\int\limits_r^{r_\infty} \Phi(y)\frac{\mathrm{d}(y^2)}{\sqrt{y^2-r^2}} = -\frac{1}{\pi}\int\limits_r^{r_\infty} \Phi^{'}(y)\frac{\mathrm{d}y}{\sqrt{y^2-r^2}}. \tag{10.11}$$

For the case of speckle photography, the parameter that has to be determined in experiments is the deflection angle of the light passing through the object, $\epsilon(y)$. So, the direct Abel equation for speckle photography is

$$\epsilon(y) = 2y\int\limits_y^{r_\infty} \frac{\partial}{\partial r}\left[\frac{n}{n_\infty}\right]\frac{\mathrm{d}r}{\sqrt{r^2-y^2}}, \tag{10.12}$$

and the inverse Abel transform is

$$\frac{n(r)-n_\infty}{n_\infty} = -\frac{1}{\pi}\int\limits_r^{r_\infty}\frac{\epsilon(y)\mathrm{d}y}{\sqrt{y^2-r^2}}. \tag{10.13}$$

Unfortunately, in experimental practice instead of the exact inverse Abel equation (10.10) we need to use an approximate equation of the type

$$\hat{f}(r) = \hat{\mathcal{A}}^{-1}\{\hat{\Phi}(y) + \beta\}, \tag{10.14}$$

where $\hat{f}(r)$ is an estimate of the solution, $\hat{\mathcal{A}}^{-1}$ is an estimate of the inverse Abel transform, $\hat{\Phi}(y)$ is incomplete and 'corrupted' experimental data, and β is the experimental noise. In the general case, optimization of the operator $\hat{\mathcal{A}}^{-1}$ is carried out by iterations with the aim of minimizing of the reconstruction error

$$\Delta_{\mathrm{rec}} = \frac{\int |\hat{f}(r) - f(r)|\,\mathrm{d}r}{\int f(r)\,\mathrm{d}r}. \tag{10.15}$$

A direct and very simple approximation for evaluating the Abel integral has been described by Schardin [403], see [12]. The test section is subdivided into N annular zones of width h: $|r_\mu - r_{\mu-1}| = h$, where $\mu = 1, ..., N$. It is assumed that the function $f(r)$ is constant within each annular zone.[2] Then the integral in Eq. (10.8) is approximated by the following summation

[2] This means that in the case of speckle photography as well as in the case of shearing interferometry the density gradient is taken as constant in each zone [12].

$$\begin{aligned}
\Phi_\mu = \Phi(y_\mu) &= 2\sum_{i=\mu}^{N} f(r_{\mathrm{i}}) \int\limits_{y_{i-1}}^{y_{\mathrm{i}}} \frac{\mathrm{d}(r^2)}{\sqrt{r^2-y_{\mathrm{i}}^2}} \\
&= 2h\sum_{i=\mu}^{N} f(r_{\mathrm{i}})\left(\sqrt{(i)^2-(\mu-1)^2}-\sqrt{(i-1)^2-(\mu-1)^2}\right) \\
&= 2h\sum_{i=\mu}^{N} f(r_{\mathrm{i}})\alpha(\mu,i).
\end{aligned} \tag{10.16}$$

The recursive solution of this system of equations is

$$f(r_\mu) = \frac{1}{2h\sqrt{2\mu-1}}\left(\Phi(y_\mu) - \sum_{i=\mu+1}^{N} \alpha(\mu,i)\cdot f(r_{\mathrm{i}})\right), \tag{10.17}$$

where

$$\alpha(\mu,i) = \sqrt{(i)^2-(\mu-1)^2} - \sqrt{(i-1)^2-(\mu-1)^2}. \tag{10.18}$$

This step-by step solution starts at the outer edge of the field with

$$f_N = \frac{\Phi_N}{2h\sqrt{2N-1}};$$

$$f_{N-1} = \frac{h}{2}\sqrt{2N-3}\left(\Phi_{N-1} - f_N\cdot(2\sqrt{N-1}-\sqrt{2N-3})\right); \tag{10.19}$$

$$\ldots$$

A related technique that uses the solution in the form of a series expansion has been successfully developed and is now referred to as the *algebraic reconstruction technique*, ART [377,404].

Leaving the further discussion of numerical reconstruction techniques for the following sections, a simple inverse Abel integral evaluation method will be briefly described here following [405]. The integral (10.10) is divided into two parts

$$f(r) = -\frac{1}{\pi}\left(\int\limits_{r}^{r+\delta} \Phi'(y)\frac{\mathrm{d}y}{\sqrt{y^2-r^2}} + \int\limits_{r+\delta}^{r_\infty} \Phi'(y)\frac{\mathrm{d}y}{\sqrt{y^2-r^2}}\right), \tag{10.20}$$

and the first integral over a small interval $[r, r+\delta]$ is evaluated using the approximation $[\Phi]'(y) = [\Phi]'(r) + (r-y)[\Phi]''(r)$

$$\int_r^{r+\delta} \Phi'(y)\frac{\mathrm{d}y}{\sqrt{y^2-r^2}} = [\Phi]'(r)\left\{\ln[(r+\delta)+\sqrt{\delta(2r+\delta)}]-\ln(x)\right\}$$
$$+[\Phi]''(r)\left(\sqrt{\delta(2r+\delta)}\right. \qquad (10.21)$$
$$\left. - r\ln[r+\delta+\sqrt{\delta(2r+\delta)}]+r\ln(r)\right).$$

This allows one to avoid the singularity at the point $y = r$ and use simple integration techniques for the evaluation of the second integral in (10.20).

The Abel inversion has been successfully used in speckle photography experiments during the last decade [21,151,394,406-409].

The experiments show that the speckle photography technique associated with the Abel transform is a reliable method of refractive index reconstruction in axisymmetrical flows. As for other optical techniques, difficulties arise in the case of flows with discontinuity surfaces, e.g. shock waves, [12]. The new possibility of speckle photography for avoiding of the 'singular' portions of flows with shocks is discussed in Section 7.3.

Let us go further to 3D flows and consider the flows with $\frac{\partial}{\partial z} << \frac{\partial}{\partial x}, \frac{\partial}{\partial y}$. For such flows we can reconstruct the distributions $f(x, y)$ for a number of z-sections, and, so, reconstruct the whole flowfield $f(x, y, z)$ by interpolation of the obtained data. For arbitrary flows integral (10.1) depends on the angular orientation of the integration path, Θ, and on the coordinate p, see Fig. 10.3:

$$\Phi(\Theta, p) = \int_{s_{\mathrm{i}}} f(x, y)\,\mathrm{d}s_{\mathrm{i}}. \qquad (10.22)$$

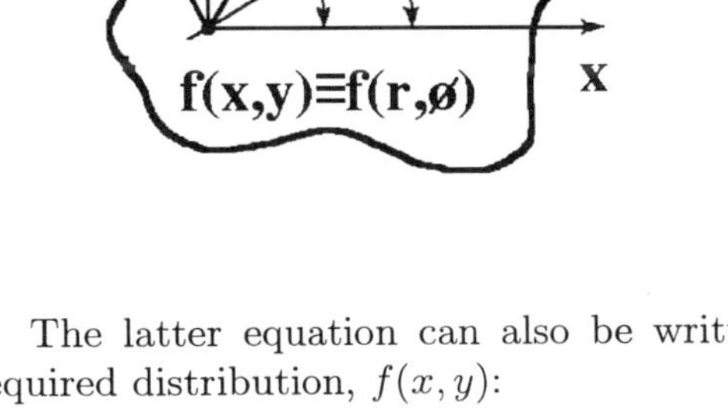

Fig. 10.3. Geometry of line-of-sight measurements for an arbitrary flow

The latter equation can also be written as a Radon transform of the required distribution, $f(x, y)$:

$$\Phi(\Theta, p) = \mathcal{R}\{f(x, y)\}. \qquad (10.23)$$

Given an infinite number of 'precise' projection measurements, the distribution $f(x, y)$ can be determined as the inverse Radon transform of the obtained data:

$$\begin{aligned} f(x,y) &= \mathcal{R}^{-1}\{\Phi(\Theta,p)\} \\ &= -\tfrac{1}{2\pi^2}\int_0^{\pi} \mathrm{d}\Theta \int_{-\infty}^{\infty} \tfrac{\Phi(\Theta,p)\,\mathrm{d}p}{(p-p_0)^2} \\ &= -\tfrac{1}{2\pi^2}\int_0^{\pi} \mathrm{d}\Theta \int_{-\infty}^{\infty} \tfrac{\Phi'(\Theta,p)\mathrm{d}p}{(p-p_0)} \\ &= -\tfrac{1}{2\pi^2}\int_0^{\pi} \mathrm{d}\Theta \int_{-\infty}^{\infty} \Phi''(\Theta,p)\ln|p-p_0|\,\mathrm{d}p, \end{aligned} \qquad (10.24)$$

where $p_0 = -x\sin\Theta + y\cos\Theta$ and Φ', Φ'' are the mean spatial derivatives in the direction of the p-coordinate, $\Phi'(\Theta,p) = \frac{\partial}{\partial p}\{\Phi(\Theta,p)\}, \Phi''(\Theta,p) = \frac{\partial^2}{\partial p^2}\{\Phi(\Theta,p)\}$.

For the case of speckle photography the latter equation has the form

$$f(x,y) = \frac{n(x,y) - n_\infty}{n_\infty} = -\frac{1}{2\pi^2}\int_0^{\pi} \mathrm{d}\Theta \int_{-\infty}^{\infty} \frac{\epsilon_p(\Theta,p)\,\mathrm{d}p}{(p-p_0)}, \qquad (10.25)$$

where ϵ_p is the probing light deflection angle along the p-axis.

In the polar coordinate system Eq. (10.24) has the following form, see [382]:

$$\begin{aligned} f(r,\phi) &= -\tfrac{1}{2\pi^2}\int_0^{\pi} \mathrm{d}\Theta \int_{-\infty}^{\infty} \tfrac{\Phi(\Theta,p)\,\mathrm{d}p}{[p - r\sin(\phi-\Theta)]^2} \\ &= \tfrac{1}{2\pi^2}\int_0^{\pi} \mathrm{d}\Theta \int_{-\infty}^{\infty} \tfrac{\Phi'(\Theta,p)\,\mathrm{d}p}{r\sin(\phi-\Theta)-p}. \end{aligned} \qquad (10.26)$$

As with the Abel transform, in the case of Radon inversion in experimental practice instead of the exact equation (10.24) we need to use the approximate equation

$$\hat{f}(x,y) = \hat{\mathcal{R}}^{-1}\{\hat{\Phi}(\Theta,p) + \beta\}. \qquad (10.27)$$

Here, in common with the case of the Abel transform, $\hat{f}$ is an estimate of f, $\hat{\mathcal{R}}^{-1}$ is an estimate of the inverse Radon transform, $\mathcal{R}^{-1}$, $\hat{\Phi}(\Theta,p)$ is incomplete and 'corrupted' experimental data, and β is the experimental noise.

Equations (10.14) and (10.27) describe so-called ill-posed mathematical problems. Strictly speaking, even the 'precise' Eqs. (10.13) and (10.24) are referred to as ill-posed mathematical problems. This means that the reconstruction procedure is very sensitive to the incompleteness and errors in the experimental data. This leads to large errors in the reconstructed distributions and to so-called 'artefacts' - numerous additional peaks in the required distributions.

Many algorithms of computerized tomography have been developed recently and some of them will be discussed in the next section.

10.2 Radon, Fourier and Hartley transforms

Let $F(\nu_x, \nu_y)$ be the Fourier transform of the required function $f(x, y)$

$$F(\nu_x, \nu_y) = \mathcal{F}\{f(x,y)\} = \int_{-\infty}^{\infty}\int_{-\infty}^{\infty} f(x,y)\exp[-2\pi\mathrm{i}(x\nu_x + y\nu_y)]\,\mathrm{d}x\mathrm{d}y. \quad (10.28)$$

This transform may be rewritten in the polar coordinate system in the Fourier plane (ν_x, ν_y)

$$F(\nu, \beta) = \mathcal{F}\{f(x,y)\} = \int_{-\infty}^{\infty}\int_{-\infty}^{\infty} f(x,y)\exp[-2\pi\mathrm{i}\nu(x\cos\beta + y\sin\beta)]\,\mathrm{d}x\mathrm{d}y \quad (10.29)$$

where

$$\nu_x = \nu\cos\beta \qquad \nu_y = \nu\sin\beta. \quad (10.30)$$

It is convenient also to define the coordinate system (p, s), rotated with respect to the (x, y) system by an angle Θ

$$p = x\cos\Theta + y\sin\Theta \qquad s = -x\sin\Theta + y\cos\Theta. \quad (10.31)$$

Here the s-axis is oriented along the probing light beam, and the p-coordinate describes the minimal distance between the coordinates centre and the light beam. The required function $f(p, s)$ in this coordinate system can be expressed through its Fourier transform, $F(\nu_p, \nu_s) = \mathcal{F}\{f(p, s)\}$:

$$f(p, s) = \int_{-\infty}^{\infty}\int_{-\infty}^{\infty} F(\nu_p, \nu_s)\exp[\mathrm{i}2\pi(p\nu_p + s\nu_s)]\,\mathrm{d}\nu_p\mathrm{d}\nu_s. \quad (10.32)$$

The line-of-sight integral from this function is

$$\begin{aligned}\Phi(\Theta,p) &= \int_{-\infty}^{\infty} f(p,s)\,\mathrm{d}s \\ &= \int_{-\infty}^{\infty}\int_{-\infty}^{\infty} F(\nu_p,\nu_s)\,\mathrm{d}\nu_p\mathrm{d}\nu_s \int_{-\infty}^{\infty} \exp[\mathrm{i}2\pi(p\nu_p+s\nu_s)]\,\mathrm{d}s.\end{aligned} \tag{10.33}$$

In the polar coordinate system Eq. (10.32) can be rewritten as

$$\begin{aligned}f(r,\phi) &= \int_0^{2\pi} \mathrm{d}\beta \int_0^{\infty} F(\nu,\beta)\exp[\mathrm{i}2\pi\nu r\sin(\phi-\beta)]\nu\,\mathrm{d}\nu \\ &= \int_0^{\pi} \mathrm{d}\Theta \int_{-\infty}^{\infty} |\nu|\Phi_\Theta(\nu)\exp[\mathrm{i}2\pi\nu p]\,\mathrm{d}\nu.\end{aligned} \tag{10.34}$$

Here, as earlier, $p = r\sin(\phi-\Theta)$.

From Eq. (10.33) we can derive an expression relating the one-dimensional Fourier transform of the projection of a function $f(p,s)$ and the slice of its two-dimensional Fourier transform

$$\mathcal{F}_1\{\Phi_\Theta(p)\}(\nu_p) = \mathcal{F}_2\{f(p,s)\}(0,\nu_p). \tag{10.35}$$

This relation is called the *Fourier slice theorem*, or the *centrer section theorem*. Using a number of probing directions, characterized by Θ, it is possible to fill up the frequency domain and the required distribution can then be reconstructed by the inverse Fourier transform

$$f(x,y) = \mathcal{F}_2^{-1}\{\mathcal{F}_1(\mathcal{R}[f]).\} \tag{10.36}$$

In practice, it is more common to use other algorithms for reconstruction, also based on the use of the Fourier slice theorem. In particular, the filtered backprojection algorithm will be described in the next section. Verhoeven [392] uses the Hartley transform instead of the Fourier transform and obtains the same relations. The advantage of doing this is that the Hartley transform of a real function is always real and numerical implementations of the Hartley transform are faster than their Fourier counterparts.

The 1D Hartley transform is expressed as

$$\mathcal{H}_1\{f(r,s)\} = \int_{-\infty}^{\infty} f(r,s)\mathrm{cas}(2\pi\rho r)\,\mathrm{d}r \tag{10.37}$$

The 1D Harley transform of a projection has the form

$$\mathcal{H}_1\{\Phi(r,\Theta)\}(\rho,\Theta) = \int_{-\infty}^{\infty}\int_{-\infty}^{\infty} f(r,s)\mathrm{cas}(2\pi\rho r)\,\mathrm{d}r\,\mathrm{d}s. \tag{10.38}$$

Changing the coordinates from (r, s) to (x, y) and making the substitution $u = -\rho \sin \Theta$ and $v = -\rho \cos \Theta$, one obtains

$$F(u,v) = \int_{-\infty}^{\infty} \int_{-\infty}^{\infty} f(x,y) \mathrm{cas}(2\pi[ux + vy]) \, \mathrm{d}x \mathrm{d}y, \qquad (10.39)$$

where $\mathrm{cas}\phi = \cos\phi + \sin\phi$. This is precisely the 2D Hartley transform of $f(x, y)$, and demonstrates the central section theorem for the Hartley–Radon pair [392]:

> The one-dimensional Hartley transform of the projections of an object at angle Θ and frequency ρ is equal to the 2D Hartley transform of the object itself at the point $u = -\rho \sin \Theta$, $v = -\rho \cos \Theta$.

10.3 Tomography algorithms

Let us again make common simplifications. First, the image is reconstructed assuming that the rays are straight. Then, the density distribution along the rays is computed and the rays traced by using the ray-equation. In the case of significant ray bending, the modified projections may be evaluated along the bent rays and then used for a second iteration to reconstruct the object. Several iterations may be needed in a strongly refractive medium.

10.3.1 Algebraic reconstruction technique, ART

The ART is the simplest approach among the various reconstruction techniques. For multi-directional tomography with a large number of projections this technique lacks accuracy and speed of implementation. However, it gains in some situations at a very low number of projections. In principle, the ART can be combined with the digital ray-tracing technique [410]. An iterative procedure for such an approach will include at least three steps. First, we construct the required distribution ignoring refraction; second, we perform ray tracing taking into account the obtained distribution; and finally these rays are used to construct a more accurate set of algebraic equations, see [411].

Let us assume that in each cell of a square grid the values of the required function $f(x, y)$ are constants, see Fig. 10.4.

Let f_m denote this constant value in the mth cell and let N be the total number of cells. A line integral will now be called a ray sum. Then for each ray the ray sum is

$$\Phi_i = \sum_{j=1}^{N} w_{ij} f_j, \qquad i = 1, 2, \ldots, M \qquad (10.40)$$

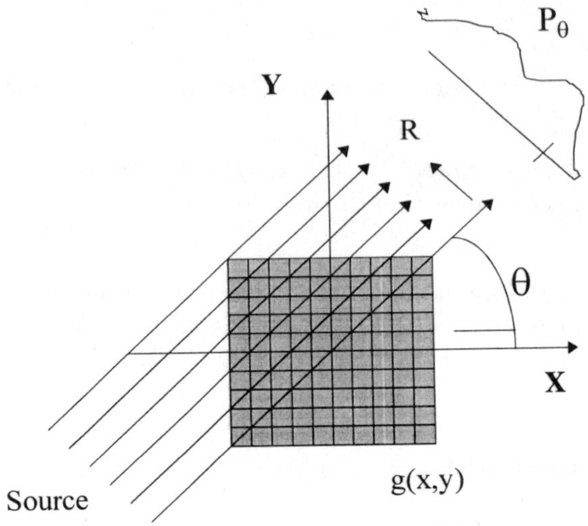

Fig. 10.4. Geometry for the ART

where M is the total number of rays in all the projections and w_{ij} is the weighting factor that represents the contribution of the jth cell to the ith ray sum.

For relatively small values of M and N, a direct inversion of matrix equation (10.40) provides the required distribution. To do this, M must be greater than N.

For large values of M and N there exist many iterative methods for solving Eq. (10.40). For instance, the following iterative procedure has been described by Kak [375]. Let us denote by $\mathbf{f}^0 = (f_1^{(0)}, f_2^{(0)}, \ldots, f_N^{(0)})$ an initial guess at the solution. Further iterations are described by

$$\mathbf{f}^{(j)} = \mathbf{f}^{(j-1)} - \frac{\mathbf{f}^{(j-1)} \cdot \mathbf{w}_\mathrm{j} - \Phi_\mathrm{j}}{\mathbf{w}_\mathrm{j} \cdot \mathbf{w}_\mathrm{j}} \mathbf{w}_\mathrm{j} \tag{10.41}$$

where $\mathbf{w}_\mathrm{j} = (w_\mathrm{j1}, w_\mathrm{j2}, \ldots, w_\mathrm{jN})$.

This equation can be rewritten taking into account that the term $(\Phi_\mathrm{j} - \Phi_\mathrm{j}^*)$ is the difference between the measured ray sum along the jth ray and the computed ray sum for the same jth ray based on the $(j-1)$th solution

$$f_m^{(j)} = f_m^{(j-1)} + \frac{\Phi_\mathrm{j} - \Phi_\mathrm{j}^*}{\sum\limits_{k=1}^{N} w_\mathrm{jk}^2} w_\mathrm{jm} \tag{10.42}$$

where

$$\Phi_\mathrm{j}^* = \mathbf{f}^{(j-1)} \cdot \mathbf{w}_\mathrm{j} = \sum_{k=1}^{N} w_\mathrm{jk}.$$

While most weight coefficients are zero, the total number of weights needed for such techniques is an enormous number. Thus even for a grid with dimensions 30×30 and 10 projections with 100 readings at each projection it is 10^6, whereas for a 100×100 grid and 100 projections it is 10^8. This is the main reason for further development of ARTs. In many practical cases the weights w_{kj} are simply replaced by 1 and 0, depending on whether the kth grid cell is within the jth ray. In this case the sum $\sum_{k=1}^{N} w_{jk}^2$ can be replaced by N_j, which is the number of cells whose centres are within the jth ray and the correction to the mth cell from the jth equation can be written as

$$f_m^{(j)} = f_m^{(j-1)} + \frac{\Phi_\mathrm{j} - \Phi_\mathrm{j}^*}{N_\mathrm{j}}. \tag{10.43}$$

As noted by Kak [411], ART reconstructions usually suffer from 'salt and pepper noise'. Many improvements of this technique were made to avoid it as well to facilitate the implementation. Among these are the SIRT and SART algorithms [375,412], the MART algorithm [413,414], the MENT algorithm [377,415], and many others.

10.3.2 Fourier transform algorithms

Here we describe algorithms based on data transformation into the Fourier or Hartley plane. With these methods projections are transformed into frequency space to obtain some of the values of the transform of the whole of the required object distribution. Other values are estimated by interpolation, and the reconstruction is performed by taking the inverse transform. The main advantages of this approach are the possibilities to use *a priori* information about the required distribution and data filtering and correction in the frequency domain.

As before, let thc Fourier transform of the required object function be

$$F(\nu_x, \nu_y) = \int_{-\infty}^{\infty}\int_{-\infty}^{\infty} f(x,y) \exp[-2\pi\mathrm{i}(\nu_x x + \nu_y y)]\, \mathrm{d}x\mathrm{d}y \tag{10.44}$$

and the inverse two-dimensional Fourier transform be

$$f(x,y) = \int_{-\infty}^{\infty}\int_{-\infty}^{\infty} F(\nu_x, \nu_y) \exp[-2\pi\mathrm{i}(\nu_x x + \nu_y y)]\, \mathrm{d}\nu_x \mathrm{d}\nu_y \tag{10.45}$$

Substituting into the projection equation in the polar coordinate system

$$\Phi_\Theta(p) = \int_0^{2\pi}\int_0^{\infty} f(r,\phi)\delta[r\cos(\Theta-\phi)-p]r\,\mathrm{d}r\mathrm{d}\phi \tag{10.46}$$

Eq. (10.45) becomes

$$F(\nu,\beta) = \int_0^{\infty} \Phi_\Theta(p)\exp[-2\pi\mathrm{i}\nu p]\,\mathrm{d}p = \mathcal{F}_1\{\Phi_\beta(p)\} \tag{10.47}$$

where β was introduced in Section 10.2 as the angle in the Fourier plane. Thus the Fourier transform of a projection at angle Θ forms a line in the Fourier plane at the same angle $\beta = \Theta$ [416].

Now the reconstruction procedure can be deduced. The projections measured in object space are line-by-line transformed into Fourier space. The whole space is filled by an interpolation of the experimental data. Then, an inverse 2D Fourier transform reveals the required distribution

$$f(x,y) = \int_0^{2\pi}\mathrm{d}\Theta\int_0^{\infty} F(\nu,\Theta)\exp[2\pi\mathrm{i}\nu(x\cos\Theta + y\sin\Theta)]\nu\,\mathrm{d}\nu. \tag{10.48}$$

Since the experimental data are discrete, the discrete form of the transform must be used. This form was first used by Bracewell [417] and Bracewell and Riddle [418] and is described by Sweeney and Vest [382] and Hunter and Collins [209].

Using the notation defined by Sweeney and Vest [382] and shown in Fig. 10.5, the discrete two-dimensional Fourier transform becomes

$$\begin{aligned} F\left(\frac{m}{L_x},\frac{n}{L_y}\right) &= \frac{1}{4B_xB_y}\sum_{l=-\infty}^{\infty}\sum_{k=-\infty}^{\infty} f\left(\frac{l}{2B_x},\frac{k}{2B_y}\right) \\ &\quad\times\exp\left[-2\pi\mathrm{i}\left(\frac{ml}{2B_xL_x}+\frac{nk}{2B_yL_y}\right)\right] \end{aligned} \tag{10.49}$$

and the discrete inverse Fourier transform is

$$\begin{aligned} f\left(\frac{l}{2B_x},\frac{k}{2B_y}\right) &= \frac{1}{L_xL_y}\sum_{m=-\infty}^{\infty}\sum_{n=-\infty}^{\infty} F\left(\frac{m}{L_x},\frac{n}{L_y}\right) \\ &\quad\times\exp\left[2\pi\mathrm{i}\left(\frac{ml}{2B_xL_x}+\frac{nk}{2B_yL_y}\right)\right] \end{aligned} \tag{10.50}$$

where L_x and L_y are the spatial limits of the function $f(x,y)$, and B_x and B_y are its bandwidth.

To determine the values of the functions outside of the known region, an extrapolation procedure is used. Thus, Sweeney and Vest [382] used the sinc-expansion technique, when values of the function $f(x,y)$ located between the sampled locations $x = \frac{l}{2B_x}$ and $y = \frac{k}{2B_y}$ are obtained by

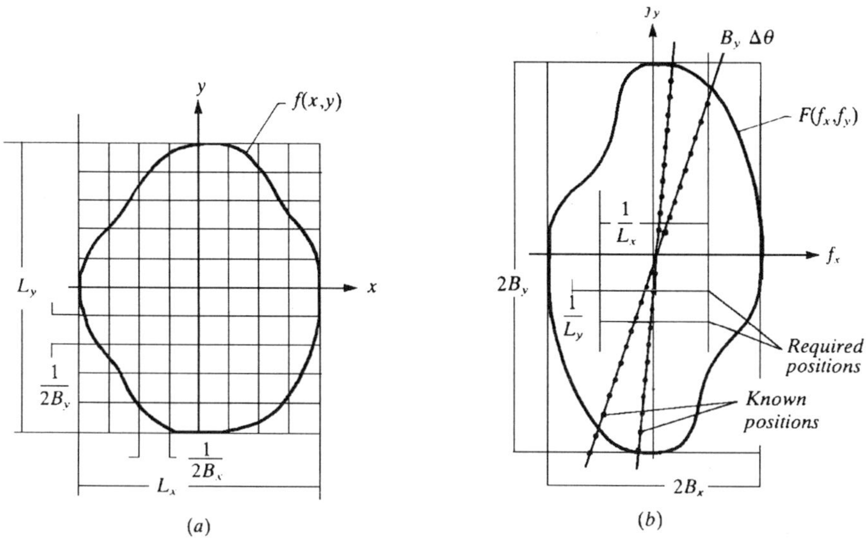

Fig. 10.5. Notation for the discrete Fourier transform

$$
\begin{aligned}
f(x,y) \quad = & \quad \sum_{l=-\infty}^{\infty} \sum_{k=-\infty}^{\infty} f\left(\tfrac{l}{2B_x}, \tfrac{k}{2B_y}\right) \\
& \times \operatorname{sinc}\left[2B_x\left(\tfrac{l}{2B_x} - x\right)\right] \cdot \operatorname{sinc}\left[2B_y\left(\tfrac{k}{2B_y} - y\right)\right] .
\end{aligned} \qquad (10.51)
$$

A similar expansion can be used in the frequency plane. The interpolation is needed because the inverse transform equation requires data from the Fourier plane over a rectangular array of points whereas the series of direct transforms provide data in the Fourier plane over a circular array, see Fig. 10.5.

Hunter and Collins [209] describe the interpolation based on the Fourier-series expansion

$$
f(x,y) = \sum_{m=-\infty}^{\infty} \sum_{n=-\infty}^{\infty} F_{mn} \exp\left\{2\pi \mathrm{i}\left[\left(\frac{mx}{L_x}\right) + \left(\frac{ny}{L_y}\right)\right]\right\}, \qquad (10.52)
$$

where

$$F_{mn} = \frac{1}{L_x L_y} \int\limits_{-\frac{L_x}{2}}^{\frac{L_x}{2}} \int\limits_{-\frac{L_y}{2}}^{\frac{L_y}{2}} f(x,y) \exp\left\{-2\pi\mathrm{i}\left[\left(\frac{mx}{L_x}\right) + \left(\frac{ny}{L_y}\right)\right]\right\} \mathrm{d}x\mathrm{d}y. \tag{10.53}$$

They have shown that the Fourier series interpolation technique in the object plane exactly corresponds to the sinc-expansion technique in the Fourier plane:

$$\begin{aligned} F(\nu_x, \nu_y) &= \sum_{m=-\infty}^{\infty} \sum_{n=-\infty}^{\infty} F\left(\frac{m}{L_x}, \frac{n}{L_y}\right) \\ &\times \mathrm{sinc}\left[L_x\left(\frac{m}{L_x} - \nu_x\right)\right] \cdot \mathrm{sinc}\left[L_y\left(\frac{n}{L_y} - \nu_y\right)\right]. \end{aligned} \tag{10.54}$$

It can be seen that the completeness of the frequency data is determined by the angle of view available. Thus for a limited angle of view the extrapolation to large areas in the Fourier plane is necessary.

Simultaneous by with the data expansion it is convenient to filter and to correct the data using *a priori* information about the object. Thus an *a priori* bandwidth constraint, defining the upper cut-off frequency of the projection data as well as a space-limiting constraint on the object in the spatial domain could be imposed. Different filters, including the Gerchberg–Papoulis, Ramachandran–Lakshminarayaanan, Shepp–Logan, and many others were developed for CT and are described in the literature [377,419,420].

The iterations are often used to introduce such a correction during the reconstruction procedure. Taking this idea further, Sato *et al.* [421,422] have developed a real-time tomographic image optical processor.

Denoting the Fourier transform of a projection at the angle Θ by $S_\Theta(\nu_\Phi)$

$$S_\Theta(\nu_\Phi) = \int\limits_{-\infty}^{\infty} \Phi_\Theta(p) \exp[-2\pi\mathrm{i}\nu_\Phi p]\, \mathrm{d}p. \tag{10.55}$$

Using the central section theorem it is possible to show that the required function $f(x,y)$ can be reconstructed by using the following two operations

$$f(x,y) = \int\limits_0^{\pi} Q_\theta(x\cos\Theta + y\sin\Theta)\, \mathrm{d}\Theta \tag{10.56}$$

where

$$Q_\theta(r) = \int\limits_{-\infty}^{\infty} S_\Theta(\nu_\Phi)|\nu_\Phi| \exp(\mathrm{i}\nu_\Phi r)\, \mathrm{d}\nu_\Phi \tag{10.57}$$

Using these formulas one first calculates a 'filtered projection' $Q_\Theta(r)$ by using Eq. (10.57), and then uses Eq. (10.56) to reconstruct the function $f(x,y)$.

10.3.3 Back projection algorithm

In this technique the reconstruction at each point of the projection is found by summing all contributions from the projections. The technique is described, e.g., in [375,377].

10.3.4 Convolution back projection algorithm

Of the transform methods, the convolution back projection (CBP) method is the most widely used. This technique is currently used in almost all applications using X-rays [411]. The method is also widely used for CT in fluid mechanical measurements, including speckle tomography; see for instance [146,387,391,416].

The back projected density due to the projection Φ_Θ at angle Θ is

$$B_\Theta(x,y) = \int_{-\infty}^{\infty} \Phi_\Theta \delta(x\cos\Theta + y\cos\Theta - p)\,\mathrm{d}p \tag{10.58}$$

The 'blurred' image is constructed by adding up the contributions at all angles

$$\begin{aligned} f_b(x,y) &= \int_0^\pi B_\Theta(x,y)\,\mathrm{d}\Theta \\ &= \int_0^\pi \mathrm{d}\Theta \int_{-\infty}^{\infty} \Phi_\Theta \delta(x\cos\Theta + y\cos\Theta - p)\,\mathrm{d}p. \end{aligned} \tag{10.59}$$

Hesselink [416] has shown that the equation for the blurred image may be transformed into

$$f_b(x,y) = \int_0^\pi \mathrm{d}\Theta \int_{-\infty}^{\infty} \frac{\mathcal{F}\{\Phi_\Theta(p)\}}{\nu} \exp[2\pi\mathrm{i}\nu(x\cos\Theta + y\cos\Theta)]|\nu|\,\mathrm{d}\nu. \tag{10.60}$$

Denoting the Fourier transform of a projection at the angle Θ by $S_\Theta(\nu_\Phi)$

$$\begin{aligned} S_\Theta(\nu_\Phi) &= \mathcal{F}_1\{\Phi_\Theta\}(p) \\ &= \int_{-\infty}^{\infty} \Phi_\Theta(p)\exp[-2\pi\mathrm{i}\nu_\Phi p]\,\mathrm{d}p. \end{aligned} \tag{10.61}$$

and using the central section theorem it is possible to show that the required distribution $f(x,y)$ can be reconstructed by using the following two operations

$$f(x,y) = \int_0^{\pi} Q_\theta \exp[2\pi \mathrm{i}(x\cos\Theta + y\sin\Theta)]\,\mathrm{d}\Theta \tag{10.62}$$

where

$$Q_\theta(p) = \int_{-\infty}^{\infty} S_\Theta(\nu_\Phi)|\nu_\Phi| \exp(2\pi \mathrm{i}\nu_\Phi p)\,\mathrm{d}\nu_\Theta \tag{10.63}$$

Comparison of Eqs. (10.62)–(10.63) and (10.60) show that the response function in the blurred image is $\frac{1}{\nu_\Theta}$ and

$$f_b(x,y) = f(x,y) \otimes \frac{1}{\nu_\Phi}. \tag{10.64}$$

So, the correction needed to be done in the Fourier plane is

$$F_b(\nu_\Phi,\beta) = \frac{F(\nu_\Phi,\beta)}{\nu_\Phi} \quad \text{or} \quad F(\nu_\Phi,\beta) = \nu_\Phi F_b(\nu_\Phi,\beta). \tag{10.65}$$

Taking it into account Eqs. (10.62)–(10.63) this may be rewritten as

$$\begin{aligned} f(x,y) &= \int_0^{\pi} \mathrm{d}\Theta \int_{-\infty}^{\infty} \frac{\mathcal{F}_1\{\Phi_\Theta(p)\}|\nu_\Phi|}{|\nu_\Phi|} \\ &\quad \times \exp[2\pi \mathrm{i}\nu_\Phi(x\cos\Theta + y\sin\Theta)]|\nu_\Phi|\,\mathrm{d}\nu_\Phi \\ &= \int_0^{\pi} \mathrm{d}\Theta \int_{-\infty}^{\infty} [\mathcal{F}_1\{\Phi_\Theta(p)\}|\nu_\Phi| \exp(2\pi\nu_\Phi p) \\ &\quad \times \delta(x\cos\Theta + y\sin\Theta - p)\,\mathrm{d}p \\ &= \int_0^{\pi} \mathrm{d}\Theta \int_{-\infty}^{\infty} \mathcal{F}_1^{-1}[\mathcal{F}_1\{\Phi_\Theta(p)\}|\nu_\Phi|] \\ &\quad \times \delta(x\cos\Theta + y\sin\Theta - p)\,\mathrm{d}p. \end{aligned} \tag{10.66}$$

The Fourier transforms in the latter expression may be replaced by the convolution

$$\mathcal{F}_1^{-1}[\mathcal{F}_1\{\Phi_\Theta(p)\}|\nu_\Phi|] = \Phi_\Theta(p) \otimes \mathcal{F}_1^{-1}\{|\nu_\Phi|\}. \tag{10.67}$$

To perform the inverse Fourier transform of $|\nu_\Phi|$ one needs to determine the frequency bandwidth. One such choice is $|\nu_\Phi|$ multiplied by a limiting rectangle function [416]

$$\mathcal{F}_1^{-1}\{|\nu_\Phi| \mathrm{rect}\frac{\nu_\Phi}{2\nu_0}\} = \nu_0[2\cdot \mathrm{sinc}(2\nu_0 p) - \mathrm{sinc}^2(\nu_0 p)]. \tag{10.68}$$

Additional filters can be designed and convoluted with this filter to further reduce the noise. However, this will result in a reduction of the spatial resolution in the reconstructed distribution [146].

Using these formulas one first calculates a 'filtered projection' $Q_\Theta(r)$ by using Eq. (10.57) and convoluting with (10.68), and then uses the back projection equation (10.56) to reconstruct the function $f(x, y)$.

The CBP algorithm was adopted for speckle photography by Lui *et al.* [146]. The required function $f(x, y)$ was determined by the equation

$$f(x,y) = \int_0^\pi \epsilon(p,\Theta) \otimes q(p)\,\mathrm{d}\Theta \tag{10.69}$$

where $\epsilon(p, \Theta)$ is the deflection angle of the light passing through object under study and $q(p)$ is a filter function. In these experiments specklegrams were taken in N different viewing directions with equal angular intervals, $\Delta\Theta$, and the specklegrams were evaluated along the p-direction at M reading points of equal spacing Δp. Thus, Eq. (10.69) in discretized form is given by

$$f(k\Delta x, n\Delta y) = \frac{1}{\pi N} \sum_{j=0}^{N-1} \sum_{m'=-\frac{M-1}{2}}^{\frac{M-1}{2}} \epsilon(m'\Delta p, j\Delta\Theta) q(m - m'), \tag{10.70}$$

where Δx and Δy are the grid sizes in the x and y directions of the discretized reconstruction plane. The filter function was chosen as

$$q(m) = \begin{cases} \frac{1}{m}, & \text{when } m \text{ is odd} \\ 0, & \text{otherwise,} \end{cases} \tag{10.71}$$

where $m = 0, \pm 1, \pm 2, \ldots, \pm \frac{M-1}{2}$.

10.3.5 Regularization-based technique

We now describe below an adaptation of these methods for speckle photography based on the paraxial approximation, when the trajectory of a separate ray of a narrow probing beam projected on to the (p, s)-plane is determined by the ray equation, see [24]

$$\frac{\mathrm{d}}{\mathrm{d}s}\left(n\frac{\mathrm{d}p}{\mathrm{d}s}\right) = \frac{\mathrm{d}n}{\mathrm{d}p} \tag{10.72}$$

considering that for the whole beam the values of the derivative $\partial/\partial p$ are constant along the p-coordinate and depend on the variable s alone, We obtain the following integral of Eq. (10.72):

$$n\frac{\mathrm{d}p}{\mathrm{d}s} - n_q\left(\frac{\mathrm{d}p}{\mathrm{d}s}\right)_q = \int\limits_{s_q}^{s}\left(\frac{\partial n}{\partial p}\right)_{s'} \mathrm{d}s'. \tag{10.73}$$

If at any point of the phase inhomogeneity the condition

$$|n(s,p) - n_0| << 1 \tag{10.74}$$

is satisfied, then n_q may be substituted for n_0 in (10.73). This enables one to integrate (10.73) and to obtain an expression for the ray trajectory:

$$p(s) = \frac{1}{n_0}\int\limits_{s_q}^{s}\int\limits_{s_q}^{s'}\left(\frac{\partial n}{\partial p}\right)_{s''} \mathrm{d}s''\mathrm{d}s' + (s - s_q)\left(\frac{\partial p}{\partial s}\right)_q. \tag{10.75}$$

The tangent to the ray trajectory in the plane of the lens L intersecting the q-plane (point A') yields a visible speckle shift due to the refraction. Use of (10.74) and (10.75) is made to get

$$\delta = \frac{1}{n_0}\int\limits_{s_q}^{s}\left(\frac{\partial n}{\partial p}\right)_s (s - s_q)\,\mathrm{d}s. \tag{10.76}$$

The equation for the ray trajectory may also be obtained, without the assumption that $\frac{\partial n}{\partial p}_s$ is constant over the beam cross-section. However, in agreement with formula (10.76), the shift will not be the same for all the rays. A point within the speckle generator when projected on to the recording plane generates a broadened spot. The picture of the appropriate variation of a speckle field cannot be obtained in the framework of geometrical optics because $\frac{\partial n}{\partial p}_s$ is not constant.

Let the required function be symmetric relative to the coordinate centre (point O). In this case, integration (10.76) arrives at

$$\delta = -\frac{s_q}{n_0}\int\limits_{s_q}^{s}\left(\frac{\partial n}{\partial p}\right)_s \mathrm{d}s. \tag{10.77}$$

As may be seen from (10.73), the quantity $\varphi = \delta/s_q$ is the angle of refraction of a ray which has passed through the phase inhomogeneity. Thus, the speckle photography method allows direct determination of the refraction angle at axisymmetric inhomogeneities. Only in this case may the refraction coefficient distribution be reconstructed using relation (10.77), as done by Farrell and Hofeldt, [17] and Wernekinck *et al.* [144].

Formula (10.76) for speckle shifts must be used for tomographic reconstruction of asymmetric distributions of the refraction coefficient.

10.3.6 Ultra low projections tomography algorithms

Given a number of line-of-sight (projection) measurements, it is possible to reconstruct the interior flowfield using one or another of the described above algorithms. Particular difficulties arise when the number of projections is extremely small. But even in the case, when this number is equal to two or three, it is possible to obtain a more general distribution than an axisymmetric distribution. For instance, we can search for a distribution in the form

$$f(r,\phi) = H(r) + K(r)\cos\phi + L(r)\sin\phi. \tag{10.78}$$

Using two orthogonal direction tomography, we can reconstruct three unknown functions, $H(r), K(r)$ and $L(r)$, with the help of the inverse Abel transform. For this, as proposed in [377,423], the following function must be calculated using the projection data $\Phi_{90}(x)$ and $\Phi_0(y)$

$$\Phi_1(x) = \frac{\Phi_{90}(x) + \Phi_{90}(-x)}{2}, \tag{10.79}$$

$$\Phi_2(x) = \frac{\Phi_{90}(x) - \Phi_{90}(-x)}{2x},$$

$$\Phi_3(y) = \frac{\Phi_0(y) + \Phi_0(-y)}{2},$$

$$\Phi_4(y) = \frac{\Phi_0(y) - \Phi_0(-y)}{2y},$$

which are the Abel transforms of the required functions

$$\Phi_1(x) = \mathcal{A}\{H(r)\}, \tag{10.80}$$

$$\Phi_2(x) = \mathcal{A}\{K(r)\},$$

$$\Phi_3(y) = \mathcal{A}\{H(r)\},$$

$$\Phi_4(y) = \mathcal{A}\{L(r)\}.$$

10.3.7 Data smoosing and interpolation

Practically all CT algorithms need data filtering, smoothing, or/and interpolation. Thus, in some algorithms it is more convenient to use data other than in the centres of a fixed grid. So, one needs to use interpolation to obtain data at any intermediate positions inside the grid cell. Thus, for the so-called *simultaneous algebraic reconstruction technique*, SART [412], Eq. (10.40) is rewritten as

$$\Phi_{\mathrm{j}} = \sum_{m=1}^{P_{\mathrm{j}}} f(s_{jm})\Delta s \tag{10.81}$$

where s_{jm} are an equispaced set of points along the jth ray as shown in Fig. 10.4. The value $f(s_{jm})$ was represented in terms of the four nearest f_n's by bilinear interpolation:

$$f(s_{jm}) \sum_{i=1}^{4} d_{jm_{\mathrm{i}}} f_{\mathrm{i}}. \tag{10.82}$$

Using this, Eq. (10.81) will have the form of Eq. (10.40) with weights

$$w_{ij} = \sum_{m=1}^{P_{\mathrm{j}}} d_{jm_{\mathrm{i}}} \Delta s, \tag{10.83}$$

where $P_{\mathrm{j}} + 2$ is the total number of points along the jth ray. It was found in [412], that setting Δs equal to half of the grid spacing δprovides a good trade-off between the accuracy of representation and computational cost.

Algorithms RICSS (Radon inversion with cubic spline smoothing) [377,424] and MART (multiplicative algebraic reconstruction technique) [392,414] use cubic B-splines for data smoothing.

Using Fourier (Hartley) transform techniques it is necessary to extrapolate data in the Fourier (Hartley) plane. Such extrapolation can be made by expanding the transform in a series of sinc-functions [425]

$$F(u,v) = \sum_{m=-\infty}^{\infty} \sum_{m=-\infty}^{\infty} F(u_m, v_n) \cdot \mathrm{sinc}[\frac{(u_m - u)}{\delta u}] \cdot \mathrm{sinc}[\frac{(v_n - v)}{\delta v}], \tag{10.84}$$

where δ_u, δ_v are the grid point spacing in the frequency plane and sinc x = $\frac{\sin \pi x}{\pi x}$.

10.4 Performance of CT algorithms

The performance of reconstruction algorithms is tested by using mathematical *phantom* objects. A phantom is a 2D mathematical function created artificially to simulate the reconstruction procedure. Starting from this function, line-of-sight *phantom* integrals (*phantom* projections) may be evaluated by the integration of *phantom* data over optical paths. In order to simulate experimental noise, a random function of definite amplitude may be added to the projections. Using one or another algorithm, an estimate of the distribution sought $\hat{f}(x, y)$ may be reconstructed from the projections corrupted by noise which may then be compared to the original object (phantom) to provide a quantitative evaluation of the reconstruction algorithm.

The reconstruction of data obtained using speckle photography projection measurement at the Belarus Heat and Mass Transfer Institute (HTMI) were performed with the help of a modified RICSS technique, called here MRICSS (modified Radon inversion with cubic spline smoothing). This algorithm was originally developed at the Siberian Branch of the Academy of Science under the guidance of Professor N.G.Preobrazhenskii and Dr. V. Pickalov for emission plasma diagnostics and was subsequently named RICSS, RICSS1 and RICSS2, see [376,377,424,426-428] and was later modified at HMTI, Minsk for speckle photography measurements [20,21,373,397,429,430]. This algorithm described in Subsection 10.3.5.

As already mentioned, the accomplishment of the Radon transformation refers to ill-posed problem of mathematical physics. This means that even for complete and absolutely precise experimental data the evaluation of the Radon integral gives rise to computational instabilities called artefacts. Additional instabilities in computing the reconstruction on are caused by the incompleteness and noise of primary experimental data. Therefore, the main points to be determined in preparing tomographic measurements are as follows:

- the best reconstruction algorithm;
- the minimum number of measurement directions needed to reconstruct the local values of the parameter under study to a specified accuracy;
- the optimum number of reading points in each direction;
- the admissible error of the readings;
- the effect of noise on the accuracy of the reconstruction.

The employment of speckle photography in a tomographic experiment is profitable regarding these points. Indeed, in this case, just as in interferometry, the functional dependence of F on the measured parameters appears to be quite simple and unique. In speckle photography, the deflection angles of the light rays are measured by the distance between Young's interference fringes, which are associated with the density gradient in the flow. The second salient feature of speckle photography is the possibility of automated reading of the light deflection angle at an arbitrary number of specklegram points in each direction. Here, the error of determining $\epsilon(\alpha, p_0)$ is not large than 1–2%.

Four different types of asymmetric functions representing phase objects were reconstructed. The main type is a multi-peaked Gaussian function including functions with 1–4 peaks. The second type is a single-peaked function with high frequency information similar to those, used by Sweeney and Vest [382] and by Bahl and Liburdy [431].

These functions are

$$f(x,y) = \sum_{i=1}^{4} a_i \exp -[b_i(x - x_{io})^2 + c_i(y - y_{i0})^2]. \tag{10.85}$$

In particular, for double-peaked functions the same parameters as in [431] were chosen:

$$\begin{aligned} f_{2-1}(x,y) &= 5\exp[-8x^2 - 25(y+0.22)^2] \\ &\quad +10\exp[-8x^2 - 25(y-0.22)^2], \end{aligned} \tag{10.86}$$

and

$$\begin{aligned} f_{2-2}(x,y) &= 5\exp[-4.5(x-0.05)^2 - 4.5(y-0.08)^2] \\ &\quad +10\exp[-22(x-0.1)^2 - 35(y-0.18)^2]. \end{aligned} \tag{10.87}$$

Cosine and *cosGauss* phantoms are also used for algorithms performance analysis [392]:

$$\begin{aligned} f_{\cos}(x,y) &= 0.25[1 - \cos 2\pi(x+0.5)^{4/5}] \\ &\quad \times[1 - \cos 2\pi(y+0.5)^{2/3}], \quad \text{for} \quad |x| < 0.5, \end{aligned} \tag{10.88}$$

and

$$f_{\cos\mathrm{Gauss}}(x,y) = 1.09\{0.3 f_{\cos}(x,y) + 0.8 f_{\mathrm{Gauss}}(x,y)\} \tag{10.89}$$

where

$$\begin{aligned} f_{\mathrm{Gauss}}(x,y) &= \exp[-81(x-0.2)^2 - 36(y-0.1)^2] \\ &\quad +\exp[-81(x+0.2)^2 - 900(y+0.35)^2]. \end{aligned}$$

An important part of the reconstruction simulation is the analysis of the experimental noise effect. Generally, the accuracy of the reconstruction procedure degrades with the added noise. This noise may be significantly amplified in the determination of the variable of interest and may even in some cases the iteration process to diverge.

The phantom noise in the simulation of the CT algorithm is characterized by the parameter δ_{noise}

$$\delta_{\text{noise}} = \frac{||\Phi_i - \Phi_i^{'}||}{||\Phi_i||} \tag{10.90}$$

where $\Phi^{'}$ are the projections associated with the noisy data. The Gaussian noise was taken to have a fixed zero mean with a variable standard deviation σ_{noise}.

Several indexes of reconstruction quality were determined to characterize different types of errors [374,392].

For a function defined and reconstructed on the grid with dimensions $M \times N$, the average error is given by

$$\Delta_{\text{a}} = \frac{\sum_{m=1}^{M} \sum_{n=1}^{N} |f(m,n) - \hat{f}(m,n)|}{MN}, \tag{10.91}$$

the *normalized* rms error is

$$\Delta_{\text{rms}} = \left\{ \frac{\sum_{m=1}^{M} \sum_{n=1}^{N} [f(m,n) - \hat{f}(m,n)]^2}{\sum_{m=1}^{M} \sum_{n=1}^{N} [f(m,n) - \overline{f}(m,n)]^2} \right\}^{\frac{1}{2}}, \tag{10.92}$$

where $\overline{f}(x,y)$ is the average value of the phantom, and the normalized absolute error is

$$\Delta_{\text{abs}} = \frac{\sum_{m=1}^{M} \sum_{n=1}^{N} |f(m,n) - \hat{f}(m,n)|}{\sum_{m=1}^{M} \sum_{n=1}^{N} f(m,n)}. \tag{10.93}$$

Herman also defined the *maximal* error of the reconstruction as follows:

$$\Delta_{\text{max}} = \max| < f_{m,n} > - < \hat{f}_{m,n} > |, \tag{10.94}$$

$$m = 1, 2, ...M - 1; \; n = 1, 2, \ldots, N - 1$$

where the functions were averaged over four values:

$$< f_{m,n} > = \frac{1}{4}(f_{m,n} + f_{m+1,n} + f_{m,n+1} + f_{m+1,n+1})$$

$$< \hat{f}_{m,n} > = \frac{1}{4}(\hat{f}_{m,n} + \hat{f}_{m+1,n} + \hat{f}_{m,n+1} + \hat{f}_{m+1,n+1}).$$

To characterize the error of the reconstruction at the location of the maximum function value, $\Delta_{f_{max}}$ is defined as

$$\Delta_{f\text{max}} = \frac{|f\text{max} - \hat{f}\text{max}|}{f\text{max}}. \tag{10.95}$$

Examples of the simulation results for speckle tomography are illustrated in Figs 10.6–10.8.

Analysing the simulation results reveals the following:

- Given a sufficient number of probing directions, it is possible to reconstruct the parameters to a prescribed accuracy in a flow of any complexity. It proves the possibility of using the integral Radon inversion for diagnostics of turbulent flows.[3]
- If a reconstruction accuracy of about 4–5% is needed, then the use of 10–13 directions is sufficient for relatively 'simple' flows with a single peak in the required parameter distribution. The number of required directions can grow somewhat when the coordinate system is chosen unsuccessfully.
- Imposing a random error within 10% on the simulated functions does not have any considerable effect on the net results when there is a sufficient number in readings at each direction of measurement and in this case the field-averaged error increases by 2–3%
- When the number of directions is small, only the centre flow region can be reconstructed to sufficient accuracy
- For the simplest flows it is sufficient to use 30–40 readings in each direction of measurement. For more complex flows the optimal number of readings increases.

10.5 Experiments

Projections can be measured in a variety of ways. Both classical and holographic interferometry as well as absorption/emission recording are the most common techniques for projection measurements in fluid mechanics experiments [381,386,391,416,432-436].

The optical path length is detected by interferometry and holography, and tomography based on such optical schemes is called phase-measurement tomography.

Speckle tomography belongs to another class of optical measurements based on recording of the deflection angle of the light passing through the flowfield. In this class, beam deflection tomography, the deflection angle may also be measured with Moiré deflectometry, shadowgraphy, or a position-sensitive detector [437-440]. Unlike to interferometry and holography, the

[3] The problem for turbulent flows is to reconstruct high-frequency vorticity. This was discussed in chapter 9.

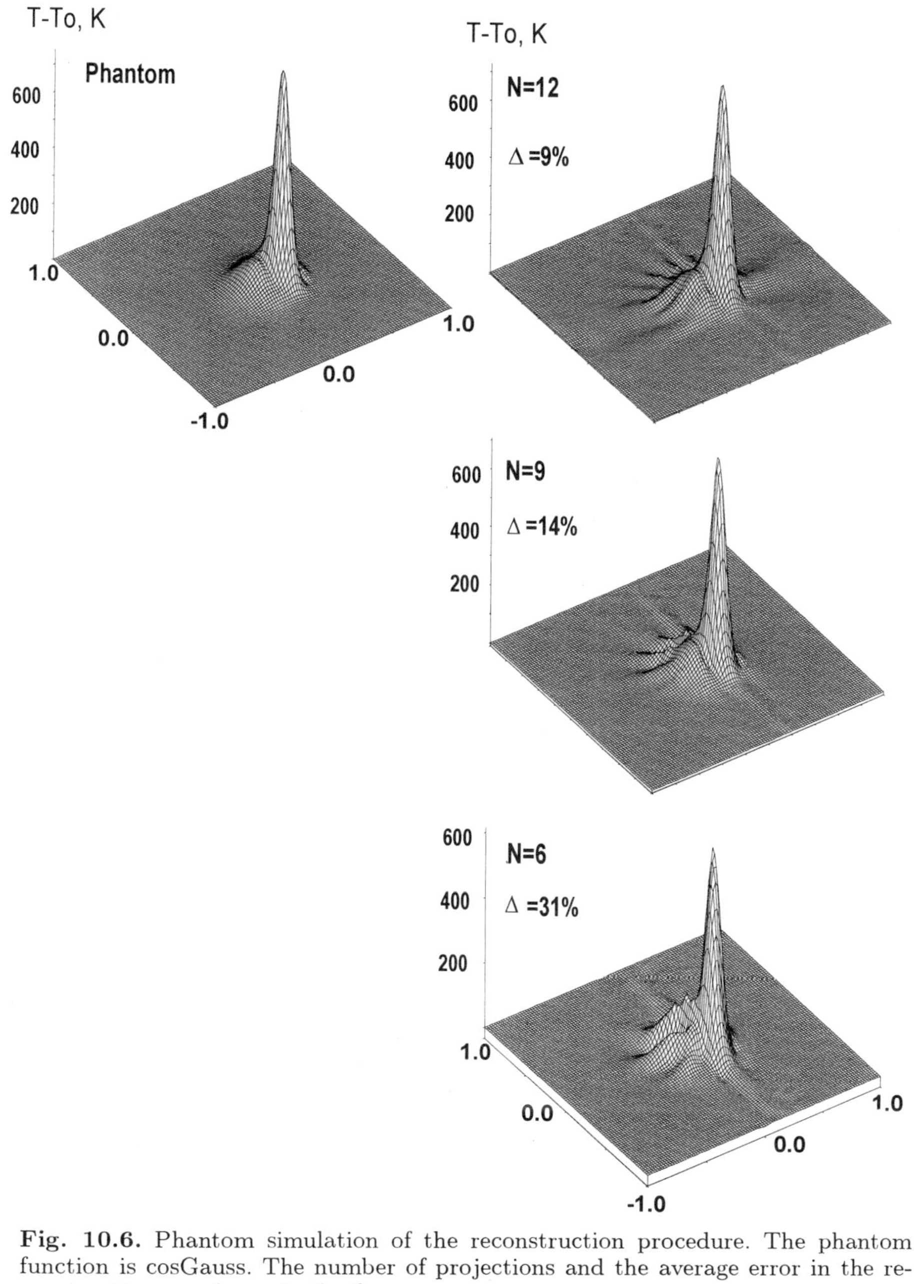

Fig. 10.6. Phantom simulation of the reconstruction procedure. The phantom function is cosGauss. The number of projections and the average error in the reconstruction are shown in the figure

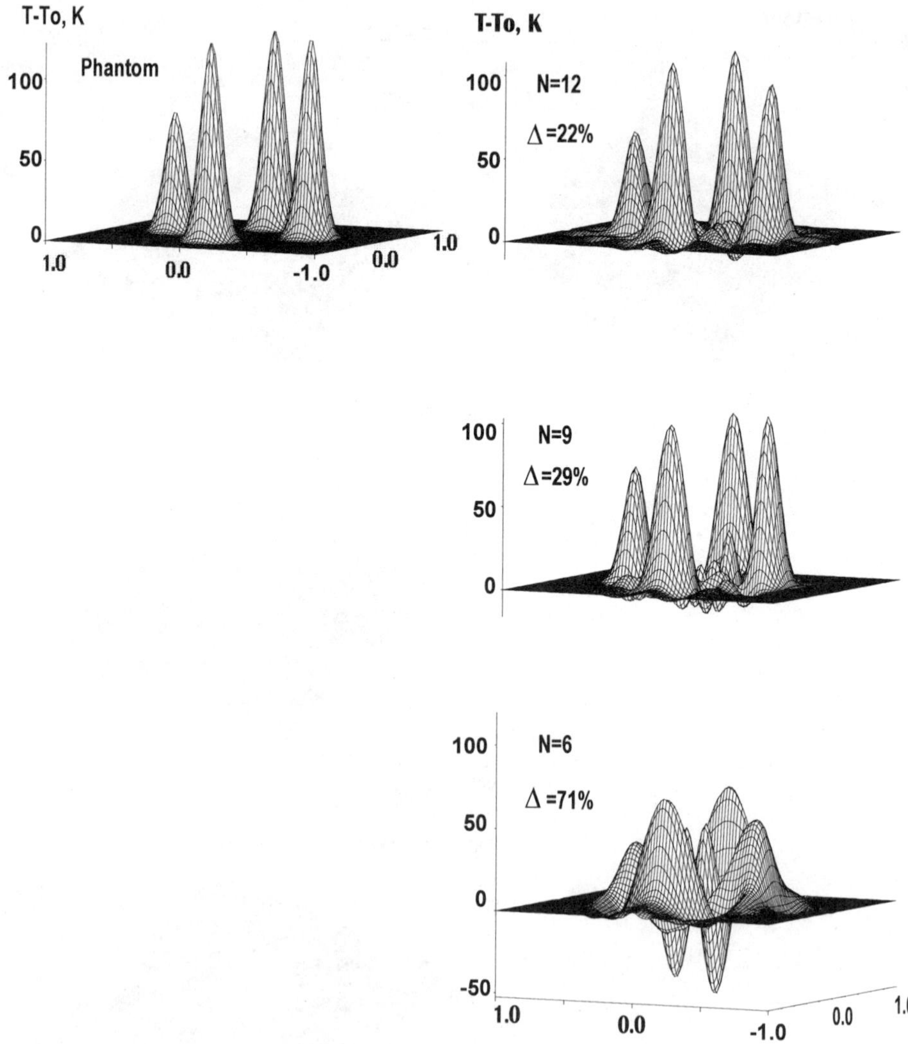

Fig. 10.7. Phantom simulation of the reconstruction procedure for a four-maxima distribution

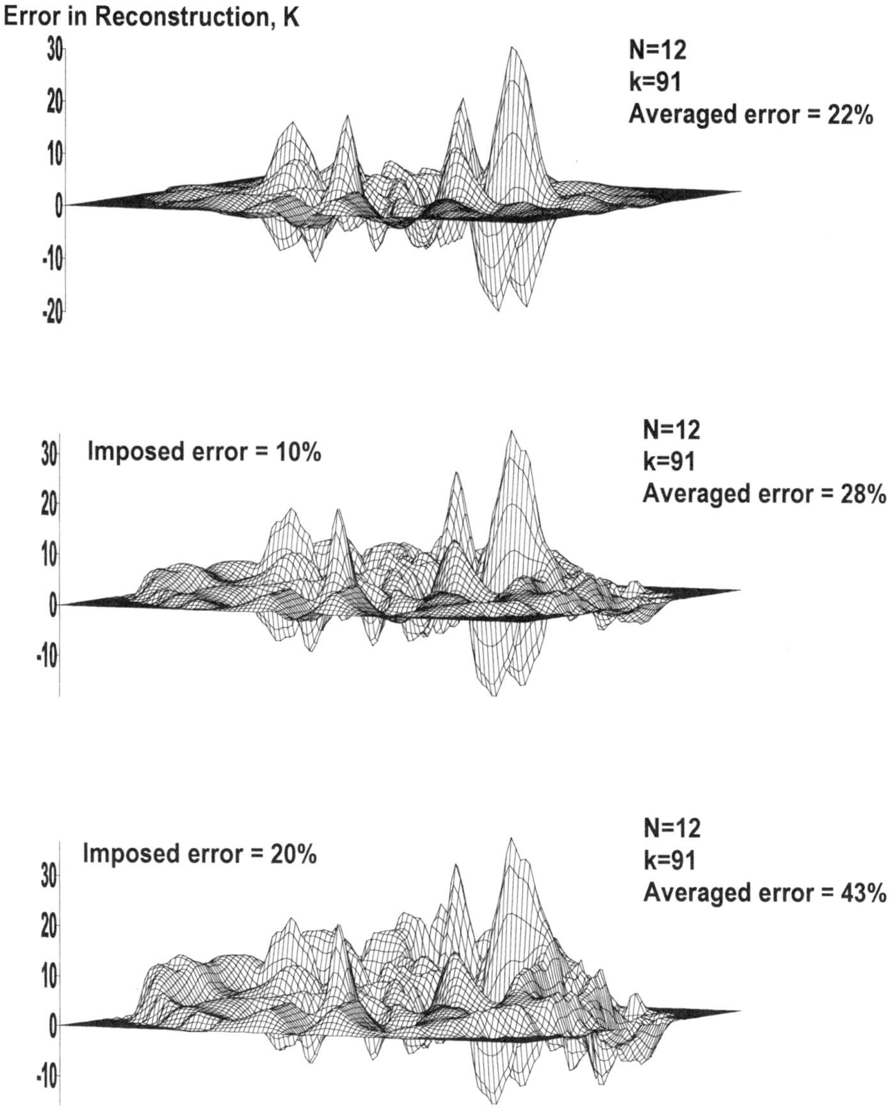

Fig. 10.8. Phantom simulation of the reconstruction procedure: influence of the imposed noise on the error in the reconstruction

beam deflection measurements require no fringe counting, which reduce ambiguity in the data and increase the dynamic range. So, the beam deflection measurements may be applied to flows where large gradients create too many fringes to resolve [438]. The deflection angle measurement does not require a reference beam, and the optical scheme for beam deflection tomography is not as complicated as that used in phase measurement tomography.

The beam is bent by refraction which is associated with the gradients of the refractive index within the flow. However, in the most cases, under reconstruction of flow parameters, refractive effects can be neglected, and CT based on the assumption of straight rays can be used.

Many physical principles of beam deflection recording were used in tomographic experiments. To show the obvious advantages of the speckle technique for beam deflection recording, let us describe some tomographic schemes, based on another methods.

Faris and Byer [438-440] have used a position-sensitive detector for deflection angle measurements. Their optical system uses a helium-neon laser focused to a 30 μm spot by a 60 mm focal length lens. Deflection angles were measured by a 78.8 mm focal length transform lens and a lateral effect position-sensitive detector. The object under study was scanned with the help of two computer controlled stepper motors. A total of 10 views with 120 reading points at each view with a horizontal spacing of 0.26 mm was taken in several planes at 0.635 mm vertical spacing above the burner. The 3D density distribution in the flowfield with dimensions approximately 10 mm $\times$ 20 mm $\times$ 20 mm was reconstructed using the obtained projections and the convolution back projection algorithm. The temperature distribution based on the air refractive index field was obtained and varied from 300 K to 2060 K. The maximum beam deflections in these experiments was 3 mrad. For such beam deflections, the paraxial approximation is valid, and deviations from straight optical paths are small.

The speckle photography technique in tomographic experiments has been used in several groups [146,151,395,397,407,409].

As already noted, speckle photography is gaining over other deflection angle measurement techniques due to its higher signal density and greater ease of computer-aided data acquisition and processing.

The optical arrangement used at Essen University is similar to that adopted by Wernekinck and Merzkirch for a spatially extended test object [145]. A camera with a shutter speed of 1/4000s was used for recording the speckle displacements during two exposures with and without flow. A beam splitter provided additional light for taking a shadowgraph of the flow pattern. A laminar helium jet was chosen as test object. The jet was imaged by a lens on to a ground glass plate, from which speckles were produced by the rays during two exposures. As in single projection speckle photography, the camera used was focused on a plane a distance l away from the ground glass and it recorded the double images of speckles, resulting from two exposures.

The speckle displacements were generated by the light deflected in the jet during the exposure with flow.

A reference displacement of speckles was superimposed by shifting the recording camera between two subsequent exposures. This shift was added to the original displacement to distinguish the signs of these measured displacements. The automatic image processing system was utilized to evaluate the specklegram obtained. The size of the reconstruction region was chosen to be 20 mm $\times$ 20 mm. Specklegrams of up to 12 viewing angles and 199 reading points of evaluation along a line on each specklegram were taken to provide input data for the tomographic reconstruction program. A cubic spline approximation for data smoothing and interpolation were performed to obtain data of 221 points in 20 mm with noise reduction. The total amount of data from 12 specklegrams was 2652 data points for one cross-section of the jet.

A convolution back projection algorithm was adopted for speckle tomography [146] and successfully applied there for the reconstruction of these data. The results show that the number of specklegrams taken in different directions affects the quality of the reconstruction. Artefacts originating from aliasing and interpolation errors become significant when the number of specklegrams used is less than eight for the axisymmetric case and 12 for the asymmetric case. The artefacts appearing in the tomograms exhibit spatial oscillations in the radial and angular directions. The angular oscillations in the outer region of the reconstruction plane are the results of angular undersampling. Their peaks and valleys form streaks, the number and direction of which are the same as the number and direction of the sampling angles. With increasing the number of angular samplings, these oscillations move away from the centre and their amplitudes decrease. The radial oscillations are due to the convolving function and the interpolation. As the number of angular samplings increases to eight for the axisymmetric case, and 12 for the asymmetric case, the oscillations become too week to be noticed [146].

Figure 10.9 contains two optical schemes for the experimental set-up of the Belarus Heat and Mass Transfer Institute.

This experimental set-up is described in publications [21,54,397]. A pulsed ruby Q-switched laser (50 mJ, 30 ns) serves as a radiation source. Laser radiation is directed across the beam splitters BS_1 and BS_2 and the mirrors M_1 and M_2 via the collimators $CL_1 - CL_4$ to the scattering plates $S_1 - S_2$, which are the speckle generators. The transparent object studied (O) can be placed both before the scattering plates, as in scheme **a**, and behind them, as in scheme **b**.

The speckle field was projected on to the photographic film by an imaging lens and recorded twice. The first exposure was made with the object under study and the second one without it. Between the two exposures the scattering plates – ground glasses – was shifted to obtain reference interferometric fringes. Four such specklegrams were obtained simultaneously (duration is

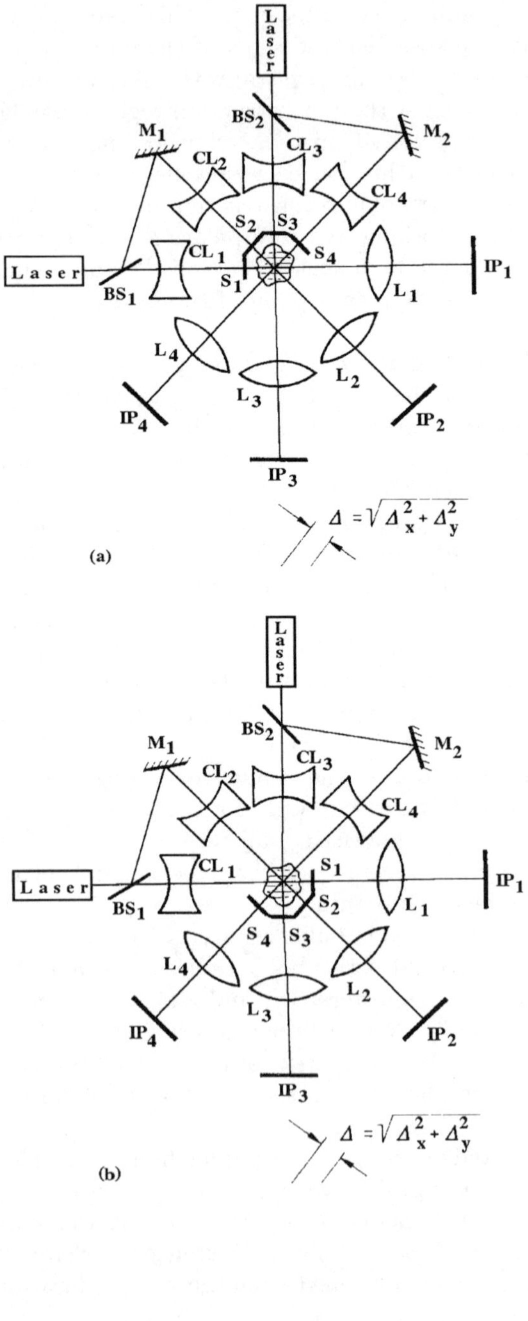

Fig. 10.9. Optical schemes for obtaining specklegrams simultaneously in four directions. (**a**) – test object (O) probing by duffuse light, and (**b**) – test object probing by collimated light. The main optical components: $S_1 - S_4$ are scattering plates (ground glasses), L_1–L_4 are imaging lenses, $CL_1 - CL_4$ are beam collimators, BS_1 and BS_2 are beam splitters, M_1 and M_2 are mirrors and IP_1 – IP_4 are image planes where the specklegrams are recorded

30 ns), and up to twelve specklegrams successively for quasi-stationary flows. The duration of the last procedure was several minutes.

Due to refraction on the test object, a distorted speckle pattern is formed when the object is present. As described earlier (see chapter 7), the magnitude of the distortion (the speckle shift at each point of specklegram) provides information about the refractive index field of the test object.

Reading and processing of the Young's fringes pattern were accomplished with the help of an automated system, the principal parts of which were described in chapter 6. The CCD array with $576\times$ 360 active elements serves as the Young's fringes pattern recorder. Image digitization was carried out using an 8-byte ADC. The resulting pattern was processed for the purpose of determining the speckle shift at each specklegram point. When the number of interference fringes was equal to about 9, the mean statistical error was no larger than 3% for the fringe contrast exceeding 20%. The scanning step along the co-ordinates x and y was specified by a computer-aided step motor, and was 100–500 μm. The beam diameter of the collimated laser used for reconstruction was equal to 500 μm.

Specklegram processing was accomplished in two steps. At the first step, a point-by-point (100 readings for each line of each direction) input of the diffraction pattern into the computer and primary processing of this pattern were realized.

As a result of this treatment, a two-dimensional array of light deflection angles was obtained and saved in computer memory. After the input of the specklegram for all directions its subsequent processing was carried out for various horizontal flowsections using an integral Radon transformation. On reconstructing the local parameters in 10–30 sections, the entire flowfield was created by interpolation of the data obtained.

Some illustrations of the speckle tomography technique will now be given.

Figure 10.10 presents the air temperature field, calculated on the basis of unidirectional speckle photography and the assumption of flow axisymmetry. The flow was the natural convection region near a vertical heated cylinder [394]. The surface temperature of the cylinder was equal to 600 K. The 3D refractive index distribution was reconstructed by successive horizontal 2D sections using integral Abel transform as described in Section 10.1.

The isobaric temperature distribution was evaluated using the Gladstone–Dale relation, which is expressed under the ideal gas assumption, as

$$n - 1 = K\rho = K\frac{p}{RT}. \tag{10.96}$$

Applying this equation once for a reference point denoted by subscript 0 and then for any arbitrary point, and then combining the two equations gives [409]

$$\frac{T(r)}{T_0} = \left[\left(\frac{n(r)}{n_0} - 1\right)\left(\frac{RT_0}{KP_0} + 1\right) + 1\right]^{-1}. \tag{10.97}$$

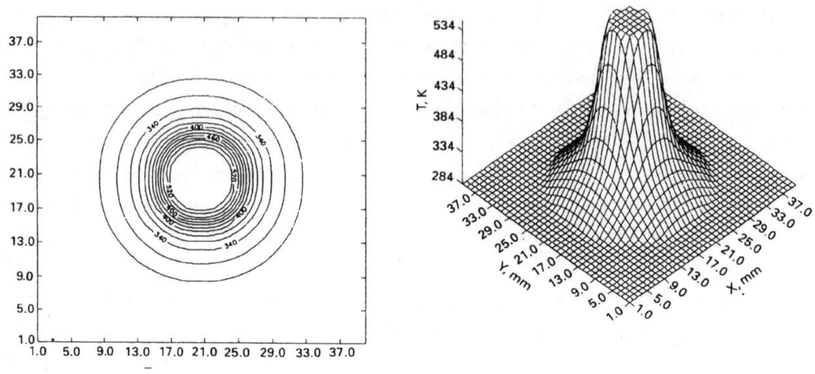

a b

Fig. 10.10. Temperature isolines (a) and profile (b) in the natural convection region near a vertical heated rod. The numbers at the isolines denote temperature (K); the coordinate axes are given in mm.

The precision of the reconstruction is higher when the properties at the reference point are specified at the centreline.

The given data (Fig. 10.10) correspond to one of the horizontal sections in the convection region, whereas a single specklegram, obtained for about 30 ns, contains the information about the entire 3D density field of the flow. Using standard holographic plates with magnification M=1, flows with a typical size of about 0.1 m may be registered. Detailed comparison of such speckle photography data with interferometric and thermocouple measurements show very good correlations [17,21,63,394,409,441].

Figure 10.11 gives the density and temperature distributions, reconstructed by using the specklegrams in an open asymmetrical flame [54]. the temperature profile in one of the vertical cross-section of this flame is shown in Fig. 10.12. These data were acquired using the Radon transformation of the deflection angles, recorded simultaneously in four directions. In the data processing, the Gladstone–Dale coefficient was assumed to be equal to that for air. Unfortunately, this coefficient is a function of the gas composition. Therefore, additional unknowns arise in combustion diagnostics.

Weinberg [8] shows that the change in composition due to the combustion of methane causes a change in the index of refraction of less than 2%. Thus, to good accuracy, the composition variation of a methane flame may be treated as the diffusion of methane and its combustion products into air [438]. The methane concentration at the position corresponding to the peak temperature is low. Therefore one can estimate the peak temperature based upon the index

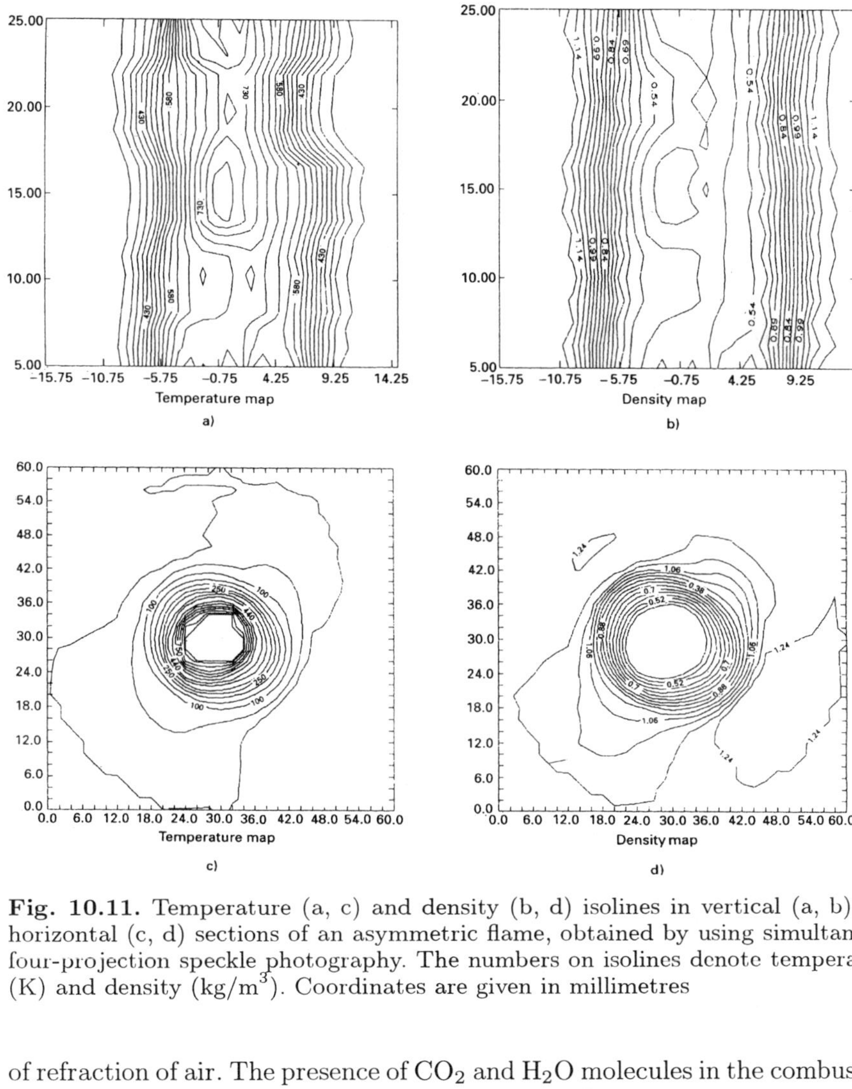

Fig. 10.11. Temperature (a, c) and density (b, d) isolines in vertical (a, b) and horizontal (c, d) sections of an asymmetric flame, obtained by using simultaneous four-projection speckle photography. The numbers on isolines denote temperature (K) and density (kg/m^3). Coordinates are given in millimetres

of refraction of air. The presence of CO_2 and H_2O molecules in the combustion products led to certain errors as well. However, the maximum concentration of these molecules in combustion products did not exceed 25%, therefore, in considering the difference of the polarizability with N_2 and O_2, the error is no larger than 10%. The same estimations may be repeated for butane-air and paraffin–air (candle) flames.

From accurate three-dimensional index of refraction measurements, the theory of combustion processes could be used to calculate both temperature and composition variations in the image. A two-wavelength technique for separation of temperature and composition effects was suggested [442].

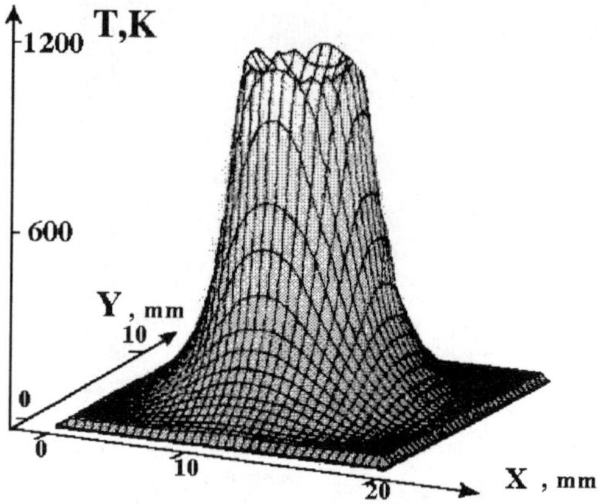

Fig. 10.12. Temperature profile in one vertical section of an asymmetric flame, obtained by using simultaneous four-projection speckle photography. Coordinates are given in millimetres

The temperature field in a narrow gap between two plane surfaces at different temperatures was studied by Blinkov *et al.* [249]. The data were obtained by processing a single specklegram, and are presented in the form of integrals of density gradients along the probing direction. It was shown that the vertical density gradients are about an order of magnitude greater than the horizontal gradients. Calculations in terms of local density values in this case are complicated due to the considerable effect of the boundary layers and the insufficient number of speckle photography views.

Achasov *et al.* have applied the speckle tomography technique to the visualization of density fields at the entrance of a supersonic turbulent underexpanded axisymmetric hot air jet into the submerged space (at ambient atmosphere) [260]. They have reconstructed the supersonic density field by using unidirectional speckle photography and the Abel transform. The jet was developed due to the efflux of air heated behind a reflected shock wavefront through a conic nozzle cut in the end wall of a low-pressure shock tube channel. The degree of nozzle expansion was 0.69, and the outlet cross-section diameter was 9.15 mm. The pressure, temperature and density behind the shock wave were calculated using an equilibrium shock adiabat in terms of the recorded propagation velocity of an incident shock wave. Velocity was measured by piezoelectric pressure gauges. The jet parameters at the nozzle cut calculated by the isentropic formulas were as follows: pressure was 3.5 atm; temperature 450 K; density 2.05 kg/m^3 and the Mach number was 1.8. Despite the fact that the values of the jet core are overestimated, speckle photography supports supersonic flow diagnostics, too.

The last example of speckle tomography reconstruction is the 3D temperature field for a four-jet combustor which is shown in Fig. 10.13 [373,430].

These data were obtained using twelve successive specklegrams and Radon inversion. It is possible to see that the obtained distribution has high spa-

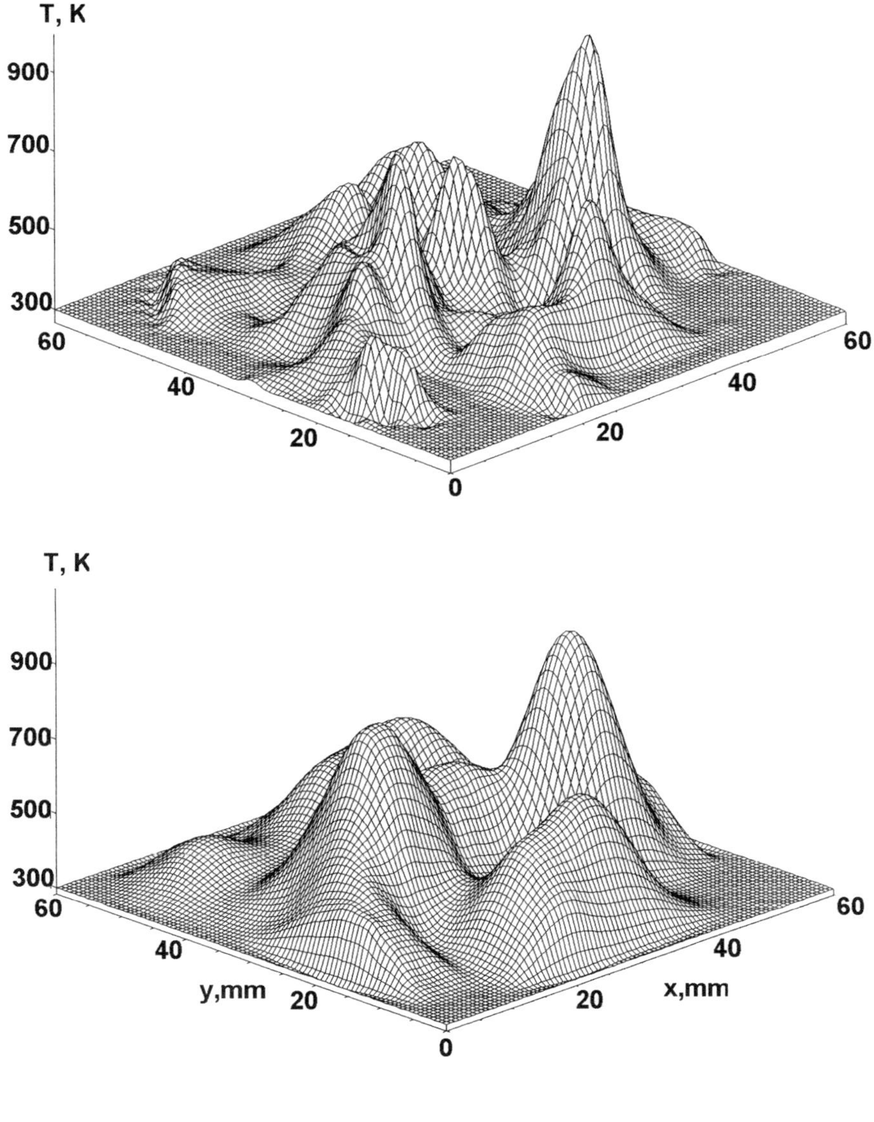

Fig. 10.13. Top, reconstructed temperature profile in one horizontal section ofa four-jet combustor; bottom, smoothed picture of the obtained distribution (12 projections)

tial frequencies, even higher than the phantom distribution. Smoothing of the obtained data led to distributions more similar to the phantom one. This indirectly proves the relatively high spatial resolution attained at these experiments, but quantitative value of this spatial resolution has to be determined in future experiments.

Thus, the presented experimental data show that the laser speckle photography technique allows one to obtain quantitative results on 3D density and temperature fields in various gas-dynamic flows including supersonic flows.

References

1. J.M. Burch and J.M.J. Tokarski. Production of multiple beam fringes from photographic scatterers. *Optica Acta*, 15(2):101–111, 1968.
2. R.K. Erf, editor. *Speckle Metrology.* Academic Press, New York, 1978.
3. M. Françon. *Laser Speckle and Applications in Optics.* Academic Press, New York, 1979.
4. R. Jones and C. Wykes. *Holographic and Speckle Interferometry.* Cambridge University Press, Cambridge, 1983.
5. J.C. Dainty, editor. *Laser Speckle and Related Phenomena*, Volume 9 of *Topics in Applied Physics.* Springer Verlag, Berlin, second enlarged edition, 1984.
6. R.S. Sirohi, editor. *Speckle Metrology.* Marcel Dekker, Inc., New York, 1993.
7. R.W. Ladenburg, editor. *Physical Measurements in Gas Dynamics and Combustion.* Princeton University Press, New Jersey, 1954.
8. F.J. Weinberg. *Optics of Flames.* Butterworth, London, 1963.
9. W. Hauf and U. Grigull. Optical methods in heat transfer. Volume 6 of *Advances in Heat Transfer*, pages 131–366. Academic Press, New York, 1970.
10. C.M. Vest. *Holographic Interferometry.* John Wiley and Sons, New York, 1979.
11. R.J. Emrich, editor. *Methods of Experimental Physics. Fluid Dynamics.* Academic Press, New-York, 1981.
12. W. Merzkirch. *Flow Visualization.* Academic Press, Orlando, second edition, 1987.
13. M. Raffel and C. Willert and J. Kompenhans *Particle Image Velocimetry. A Practical Guide*, Volume 1 of *Experiments in Fluids.* Springer Verlag, Berlin, 1998.
14. U. Köpf. Application of speckling for measuring the deflection of laser light by phase objects. *Optics Communications*, 5(5):347–350, 1972.
15. U. Köpf. Darstellung von phasenobjekten durch kohärent-optische filterung von laser-granulationsphotographiren. *Optik*, 5:592–595, 1972.
16. S. Debrus, M. Françon, C.P. Grover, M. May, and M.L. Roblin. Ground glass differential interferometer. *Applied Optics*, 11(4):853–857, 1972.
17. S. Mallick and M.L. Roblin. Speckle pattern interferometry applied to the study of phase objects. *Optics Communications*, 6(1):45–49, 1972.
18. P.V. Farrell and D.L. Hofeldt. Temperature measurement in gases using speckle photography. *Applied Optics*, 23(7):1055–1059, 1984.
19. W. Merzkirch. Density-sensitive whole-field flow measurement by optical speckle photography. *Experimental Thermal and Fluid Science*, 10:435–443, 1995.
20. N.A. Fomin. Speckle diagnostics of turbulent flows. *Optical Diagnostics in Engineering. An Electronic Journal: www.civ.hw.ac.uk./research/flic/ode.odemain.hmt*, 2(1):1–12, 1997.
21. N.A. Fomin. *Speckle Photography of Gasdynamic Flows (in Russian).* Nauka i Technica Press, Minsk, 1989.

22. J.W. Goodman. Statistical properties of laser speckle patterns. In J.C.Dainty, editor, *Speckle and related phenomena*, Volume 9 of *Topics in Applied Physics*, chapter 2, pages 9–75. Springer Verlag, Berlin, second enlarged edition, 1984.
23. A.E. Ennos. Speckle interferometry. In J.C.Dainty, editor, *Speckle and related phenomena*, Volume 9 of *Topics in Applied Physics*, chapter 6, pages 203–253. Springer Verlag, Berlin, second enlarged edition, 1984.
24. M. Born and E. Wolf. *Principles of Optics.* Academic Press, Oxford, sixth edition, 1980.
25. J.W. Goodman. *Introduction to Fourier Optics.* McGraw-Hill, San Francisco, 1968.
26. S.F. Clifford. Classical theory of wave propagation through turbulence. In J.W. Strohbehn, editor, *Laser Beam Propagation in the Atmosphere*, Volume 25 of *Topics in Applied Physics*, chapter 2, pages 18–60. Springer Verlag, Berlin, 1978.
27. L.A. Chernov. *Wave Propagation in a Random Medium.* Dover Publications, Inc., New York, 1967.
28. V.I. Tatarskii. *Wave Propagation in a Turbulent Medium.* McGraw-Hill, New York, 1961.
29. G. Adamovsky. Laser beam propagation through inhomogeneous media with shock like profiles: modeling and computing. In S.S. Cha, J.D. Trolinger, and M. Kawahashi, editors, *Optical technology in fluid, thermal and combustion flow III (SPIE - Proc. v.3172).* SPIE Optical Engineering Press, Bellingham, 1997.
30. L. Hesselink. Optical processing of images. In W.-J. Yang editor, *Handbook of Flow Visualization*, chapter 17, pages 261–278. Hemisphere Publ. Corp., Washington, 1989.
31. F.T.S. Yu. *Introduction to Diffraction, Information Processing, and Holography.* The MIT Press, Cambridge, 1973.
32. A.A. Akaev and S.A. Maiorov. *Optical Methods for Information Processing (In Russian).* Vuschaja Sckola, Moscow, 1988.
33. R.P. Khetan and F.P. Chiang. Strain analysis by one-beam laser speckle interferometry. 1: Single aperture method. *Applied Optics*, 15(9):2205–2215, 1976.
34. T. Asakura, H. Fujii, and K. Murata. Measurements of spatial coherence using speckle pattern. *Optica Acta*, 19(4):273–290, 1972.
35. T. Asakura and N. Takai. Dynamic laser speckles and their application to velocity measurements of the diffuse object. *Applied Physics*, 25:179–194, 1981.
36. V.V. Anisimov, S.M. Kosel, and G.R. Lokshin. Space-time statistical properties of coherent radiation scattered by moving diffuse object (in Russian). *Optica i Spectroscopia (USSR)*, 27:258–296, 1969.
37. T. Yoshimura. Statistical properties of dynamic speckles. *Journal of the Optical Society of America*, 3A:1032–1054, 1986.
38. J. Ohtsubo and T. Asakura. Velocity measurement of a diffuse object by using time-varying speckles. *Journal of Optics and Quantum Electronics*, 8(1):523 – 529, 1976.
39. T. Asakura. Surface roughness measurements. In R.K.Erf, editor, *Speckle Metrology*, pages 11–49. Academic Press, New York, 1978.
40. L.G. Shirley and N. George. Diffuser radiation patterns over a large dynamic range. 1: Strong diffusers. *Journal of Applied Optics*, 27(9):1850–1861, 1988.
41. J.D. Briers. Surface roughness evaluation. In R.S. Sirohi, editor, *Speckle Metrology*, pages 373–425. Marcel Dekker, Inc., New York, 1993.
42. P.F. Gray. A method of forming optical diffusers of simple known statistical properties. *Optica Acta*, 25:765–775, 1978.
43. K.D. Bonin and M.A. Kadar-Kallen. Simple diffuser for production of laser speckle. *Applied Optics*, 28(24):5293–5297, 1989.

44. H. Fujii and T. Asakura. Roughness measurements of metal surface using laser speckle. *Journal of Optical Society of America*, 67(9):1171–1176, 1977.
45. G. Parry. Speckle patterns in partially coherent light. In J.C.Dainty, editor, *Speckle and related phenomena*, Volume 9 of *Topics in Applied Physics*, chapter 3, pages 77–122. Springer Verlag, Berlin, second enlarged edition, 1984.
46. R. Meynart. Equal velocity fringes in a Rayleigh–Bénard flow by a speckle method. *Applied Optics*, 19:1385–1386, 1980.
47. R. Meynart, P.G. Simpkins, and T.D. Dudderar. Speckle measurements of convection in a liquid cooled from above. *Journal of Fluid Mechanics*, 182:235–254, 1987.
48. M.P. Arroyo, T. Yonte, M. Quintanilla, and J.M. Savirón. Particle image velocimetry in Reyleigh–Bénard convection: Photographs with high number of exposures. *Optics and Lasers in Engineering*, 9:295–316, 1988.
49. M.P. Arroyo, T. Yonte, M. Quintanilla, and J.M. Savirón. Velocity measurements in convective flows by Particle Image Velocimetry using a low power laser. *Optical Engineering*, 27(8):641–649, 1988.
50. M.P. Arroyo, T. Yonte, M. Quintanilla, and J.M. Savirón. Experimental aspects in particle image velocimetry. *Optica Pura y Aplicada*, 21:193–205, 1988.
51. M.P. Arroyo, M. Quintanilla, and J.M. Saviron. Three-dimensional study of the Rayleigh–Bénard convection by particle image velocimetry measurements. In *Proc. of ICALEO'88, LIA vol.67*, pages 187–195, Toledo, OH, 1988. Laser Institute of America.
52. A. Vogel and W. Lauterborn. Time-resolved particle image velocimetry used in the investigation of cavitation bubble dynamics. *Applied Optics*, 27:1869–1876, 1988.
53. O. Achasov, N. Fomin, O. Penyazkov, A. Oznobishin, and F. Fisson. Interferometric study of combustion in a spark ignition engine. In *Proc. of the Internationa Symposium on Internal Combustion Engines KONES'93*, pages 553–562, Gdansk-Jurata, 1993.
54. G.N. Blinkov, N.A. Fomin, and D.E. Vitkin. Speckle tomography of gas flows. In *Atlas of Visualization*, Volume 1, chapter 11, pages 171–182. Pergamon Press, Oxford, 1993.
55. N.A. Fomin, D. Vitkin, A. Krauklis, C. Gray, and C. Greated. Turbulent flowfield determination using particle image velocimetry (PIV). In *Proc. of the Minsk International Forum on Heat and Mass Transfer*, Volume 4, pages 203 –209. Heat and Mass Transfer Institute Press, Minsk, 1996.
56. O.J. Lokberg and O. Kwom. Electronic speckle pattern interferometry using a CO_2 laser. *Optics and Lasers Technology*, 16:187–192, 1984.
57. O.A. Achasov, V.G. Belkin, G.N. Blinkov, P.L. Kukharchik, A.S. Skirko, and N.A. Fomin. Infrared speckle photography on a metal thermoplastic carrier. *Optics Communications*, 68(3):171–174, 1988.
58. Y. Zhang, K.N.C. Bray, and B. Rogg. Time resolved two-dimensional visualization of flame and wall interactions. In *IMexE Conference Transaction. International Seminar on Optical Methods and Data Processing in Heat and Fluid Flows*, pages 111–116. IMechE Press, London, 1996.
59. C. Wykes and M. Flanagan. The use of a diode laser in a ESPI system. *Optics and Lasers Technology*, 19(1):37–39, 1987.
60. C.-J. Chen, Y.-G. Kim, and J.A. Walter. Progress in quantitative flow visualization and imaging process. In *Atlas of Visualization*, Volume 1, chapter 11, pages 279–295. Pergamon Press, Oxford, 1993.
61. L. Lourenço and A. Krothapalli. The role of photographic parameters in laser speckle or particle image displacement velocimetry. *Experiments in Fluids*, 5:29–32, 1987.

62. L. Lourenço. Velocity bias technique for particle image velocimetry measurement of high speed flows. *Applied Optics*, 32:2159–2162, 1993.
63. U. Wernekinck. *Anwendung der Speckle-Photographie zur Sichtbarmachung und Messung von Strőmungen mit verǎnderlicher Dichte.* PhD dissertation, Essen University, 1985.
64. I. Grant and G.H. Smith. Modern developments in particle image velocimetry. *Optics and Lasers in Engineering*, 9:245–264, 1988.
65. S. Hilgers. *Darstellung der fluidmechanischen Prozesse in Blasensäulen Reactoren durch die Particle-Image velocimetry.* PhD thesis, Essen University, 1996.
66. R.I. Soloukhin and N.A. Fomin. *Mixing Gasdynamic Lasers (in Russian).* Nauka i Technica Press, Minsk, 1984.
67. C.J.D. Pickering and N.A. Halliwell. Laser speckle photography and particle image velocimetry: photographic film noise. *Applied Optics*, 23(17):2961–2969, 1984.
68. C.J.D. Pickering and N.A. Halliwell. Particle image velocimetry: improving fringe signal-to-noise ratio with a two-step photographic process. *Journal of Optical Society of America*, 2A:610–615, 1985.
69. C.J.D. Pickering and N.A. Halliwell. Particle image velocimetry: improving fringe signal-to-noise ratio with a two-step photographic process. Part II. *Journal of Optical Society of America*, 2A:1721–1724, 1985.
70. K. Banke, K. Wutzke, and H. Helmers. Particle image velocimetry (PIV): use of photo-thermoplastic film for photographic recording of particle images. *Experiments in Fluids*, 18:136–139, 1995.
71. E.M. Barkhudarov, V.P. Berezovsky, M.I. Taktakishvili, and T.Ya. Chelidze. Infrared materials. In R.I. Soloukhin, editor, *Infrared Diagnostics Methods*, pages 52–102. Heat and Mass Transfer Institute Press, Minsk, 1982.
72. J.C. Urbach. Thermoplastic hologram recording. In H.M.Smith, editor, *Holographic Recording Materials*, pages 161–207. Springer Verlag, Berlin, 1977.
73. J.D.E. Beynon and D.R. Lamb. *Charge Coupled Devices and Their Applications.* McGraw-Hill, Maidenhead, England, 1980.
74. G.M. Carlomagno and A. Rapillo. *Handbook of Flow Visualization*, section Digital Processing of Interferograms, pages 235–260. Hemisphere Publ. Corp., New York, 1989.
75. W. Heckmann. *Auswertealgorithmen in der Particle-Image-Velocimetry.* PhD thesis, Universität GH Essen, 1995.
76. J.N. Butters and J.A. Leendertz. Holographic and videotechniques applied to engineering measurements. *Journal of Measurement and Control*, 4:349–354, 1971.
77. A. Macovski, S.D. Ramsey, and L.F. Schaefer. Time lapse interferometry and contouring using television systems. *Applied Optics*, 10:2722–2727, 1971.
78. J.N. Butters, R. Jones, and C. Wykes. Electronic speckle pattern interferometry. In R.K. Erf, editor, *Speckle Metrology*, Quantum Electronics – Principles and Applications. Academic Press, New York, 1978.
79. O.J. Løkberg. Electronic speckle pattern interferometry and its applications in rock mechanics. In S. Takemoto, editor, *Laser Holography in Geophysics*, pages 168–196. Ellis Horwood/John Wiley, New York, 1989.
80. O.J. Løkberg. Recent development in video speckle interferometry. In R.S. Sirohi, editor, *Speckle Metrology*, pages 157–195. Marcel Dekker, Inc., New York, 1993.
81. R.G. Racca, O. Stephenson, and R.M. Clements. High speed video analysis system using multiple shuttered charge coupled device imager and digital storage. *Optical Engineering*, 31(6):1369–1374, 1992.

82. M. Raffel, J. Kompenhans, B. Stasicki, B. Bretthauer, and G.E.A. Meier. Velocity measurement of compressible air flow utilizing a high-speed video camera. *Experiments in Fluids*, 18:204–206, 1995.
83. W. Merzkirch and T. Wagner. PIV with two synchronized video cameras. In *IMexE Conference Transactions. International Seminar on Optical Methods and Data Processing in Heat and Fluid Flows*, pages 237–242. IMechE Press, London, 1996.
84. T.P. Dewhirst, M.L. Jakobsen, and C.A. Greated. Multiple CCD camera DPIV for force and acceleration measurements in high-speed flows. In *IMexE Conference Transactions. International Seminar on Optical Methods and Data Processing in Heat and Fluid Flows*, pages 251–258, London, 1996. Institute of Mechanical Engineers.
85. H.T. Huang and H.E. Fielder. Reducing time interval between successive exposures in video PIV. *Experiments in Fluids*, 17:356–363, 1994.
86. B. Lecordier, M. Mouqallid, S. Vottier, E. Rouland, D. Allano, and M. Trinite. CCD recording method for cross-correlation PIV development in unstationary high speed flows. *Experiments in Fluids*, 17:205–208, 1994.
87. C.E. Willert and M. Gharib. Digital particle image velocimetry. *Experiments in Fluids*, 10:181–193, 1991.
88. W. Merzkirch, T. Mrosewski, and H. Wintrich. Digital particle image velocimetry applied to a natural convective flow. *Acta Mech.*, 4:16–26, 1994.
89. Y.Y. Hung. Displacement and strain measurement. In R.K. Erf, editor, *Speckle Metrology*, Quantum Electronics - Principles and Applications, pages 51–71. Academic Press, New York, 1978.
90. E. Archbold, J.M. Burch, and A.E. Ennos. Recording of in-plane surface displacement by double-exposure speckle photography. *Optica Acta*, 17(12):883–898, 1970.
91. E. Archbold and A.E. Ennos. Displacement measurement from double-exposure laser photographs. *Optica Acta*, 19:253–271, 1972.
92. D.E. Duffy. Measurement surface displacement normal to the line of sight. *Experimental Mechanics*, 14:378–384, 1974.
93. K.A. Stetson. A review of speckle photography and interferometry. *Optical Engineering*, 14:482–489, 1975.
94. D.W. Li, J.B. Chen, and F.P. Chiang. Statistical analysis of one-beam subjective laser-speckle interferometry. *Journal of the Optical Society of America*, 2A(5):657–666, 1985.
95. R.S. Sirohi. *Selected Paper on Speckle Metrology*. SPIE Milestone series. SPIE Optical Engineering Press, Washington, D.C., 1991.
96. R.K. Erf. Application of laser speckle to measurements. In J. W.Goodman and M. Ross, editors, *Laser Applications*, Volume 4, pages 1–69. Academic Press, New York, 1980.
97. D.E. Duffy. Moiré gauging of in-plane displacement using double aperture imaging. *Applied Optics*, 11:1778–1797, 1972.
98. B. Eliasson and F.M. Mottier. Determination of the granual radiance distribution of a diffuser and its use for vibration analysis. *Journal of the Optical Society of America*, 61:559, 1971.
99. E. Archbold, A.E. Ennos, and P.A. Taylor. A laser speckle interferometer for the detection of surface movement and vibration. In J.H. Dickson, editor, *Proc. ICO , Reading 1969*, Volume 8, Oriel Press, Newcastle-upon-Tyne, 1970.
100. L. Ek and N.E. Molin. Detection of the nodal lines and the amplitude of vibration by speckle interferometer. *Optics Communications* , 2:419–423, 1971.
101. H.J. Tiziani. Application of speckling for in-plane vibration analysis. *Optica Acta*, 18(12):891–902, 1971.

102. H.J. Tiziani. Vibration analysis and deformation measurement. In R.K. Erf, editor, *Speckle Metrology*, Quantum Electronics - Principles and Applications, pages 73–110. Academic Press, New York, 1978.
103. B. Eliasson and R. Dändliker. Contrast of photographically recorded speckle patterns. *Optica Acta*, 23:813, 1976.
104. A.F. Fercher and J.D. Briers. Flow visualization by means of single exposure speckle photography. *Optics Communications*, 37:326–330, 1981.
105. J.D. Briers. Wavelength dependence of intensity fluctuations in laser speckle pattern from biological specimens. *Optics Communications*, 13:324–326, 1975.
106. J.D. Briers and A.F. Fercher. Laser speckle technique for the visualization of retinal blood flow. In Max Born, editor, *SPIE Proc.*, Volume 369, page 22. SPIE Press, 1982.
107. J.B. Briers. Optical filtering techniques to enhance speckle contrast variations in single-exposure laser speckle photography. *Optik*, 63(3):265–276, 1983.
108. J.B. Briers and S. Webster. Quasi real-time digital version of single-exposure speckle photography for full-field monitoring of velocity or flow fields. *Optics Communications*, 116:36–42, 1995.
109. C.S. Narayanamurthy. Measurement of angular velocity using speckle photography. *Applied Optics*, 30(22):3197–3199, 1991.
110. N. Barakat, H. El-Ghandoor, and A.M. Hamed. Single exposure speckle photography applied to a slow flow field. *Optics and Laser Technology*, 21(5):328–330, 1989.
111. N.A. Fomin, D. Ragozin, H. Wintrich, and W. Merzkirch. Optical speckle patterns generated by turbulent density fields. In *Proc. of the Seminar on Optical Methods and Data Processing in Heat and Fluid Flow*, pages 211 –215, London, 1992. City University Press.
112. N.A. Fomin, W. Merzkirch, and H. Wintrich. Turbulence anisotropy measurements after shock waves by one-exposure speckle photography. In K. Takajama, editor, *Shock Waves. Proc. of Japanese National Shock Wave Symp.*, pages 455–458, Sendai, 1993. Tohoku University Press.
113. N.A. Fomin, W. Merzkirch, H. Wintrich, and D. Vitkin. Speckle photography of temperature field microstructure in a turbulent flame. In J.P.Crowder, editor, *Flow Visualization VII. Proc. of the Seventh International Simposium on Flow Visualization*, pages 904–909, New-York, 1995. Begell House, Inc.
114. N.A. Fomin, W. Merzkirch, D. Vitkin, and H. Wintrich. Visualization of turbulence anisotropy by single exposure speckle photography. *Experiment in Fluids*, 20:476–479, 1996.
115. C. Joenathan, S.M. Blair, and A.R. Ganesan. Pulsed lasers in speckle photography: error owing to pulse width. *Applied Optics*, 32(2):204–209, 1993.
116. N.A. Fomin. Speckle frame- and streak-cinematography: an insight into the structure and development of turbulence. *Archivum Combustionis*, 15(3–4):227–239, 1995.
117. N.A. Fomin and P. Khramtsov. Speckle frame and streak cinematography of temperature field microstructure in non-steady turbulent fields. In *Optical Methods and Data Processing in Heat and Fluid Flows. Proc. of International Seminar*, pages 489–496, London, 1996. Institute of Mechanical Engineers.
118. Y.Y. Hung, C.P. Hu, and C.E. Taylor. Speckle-Moiré interferometry — A tool for complete measurement of in-plane surface displacement. In *Proc. of the 7th Southeastern Conf. on Theor. Appl. Mech.,*, page 497, 1974.
119. C. Forno. White light speckle photography for measuring deformation, strain, and shape. *Optics and Laser Technology*, 16:217–221, 1975.
120. P. Boone and L.C. De Backer. Speckle method using photography and reconstruction in incoherent light. *Optik*, 44:343–355, 1976.

121. A.K.Asundi. White light in speckle metrology. In R.S. Sirohi, editor, *Speckle Metrology*, pages 325–371. Marcel Dekker, Inc., New York, 1993.
122. F.P. Chiang and A. Asundi. White light speckle method of experimental strain analysis. *Applied Optics*, 18:409–411, 1979.
123. F.P. Chiang and A. Asundi. Interior displacement and strain measurement using white light speckles. *Applied Optics*, 19:2252–2256, 1980.
124. A. Asundi and F.P. Chiang. Theory and application of the white light speckle method for strain analysis. *Optical Engineering*, 21:570–580, 1982.
125. E. Conley and G. Cloud. Practical application of double exposure non coherent light speckle photography. *Applied Optics*, 25:2246–2248, 1986.
126. E. Conley and G. Cloud. Resolution experiments using white light speckle method. *Applied Optics*, 30:795–800, 1991.
127. A. Asundi and F.P. Chiang. Applications of the white light speckle to interior displacement measurement. *Journal of Strain Analysis*, 18:23–26, 1983.
128. G. Cloud, R. Falco, R. Radke, and J. Peiffer. Non-coherent light photography for measurement of fluid velocity field. In *SPIE Proc.*, Volume 243, page 150, 1980.
129. E. Bernabeu, J.C. Amaré, and M.P. Arroyo. White-light speckle method of measurement of flow velocity distribution. *Applied Optics*, 21(14):2583–2586, 1982.
130. M. Suzuki, K. Hosoi, S. Toyooka, and M. Kawahashi. White-light speckle method for obtaining an equi-velocity map of a whole flow field. *Experiments in Fluids*, 1:79–81, 1983.
131. G.H. Kaufmann. Double pulsed white-light speckle photography. *Applied Optics*, 23:194–196, 1984.
132. M. Grobel and W. Merzkirch. White-light speckle velocimetry applied to plane free convective flow. *Experimental Heat Transfer*, 4:253–262, 1991.
133. P.T. Tokumaru and P.E. Dimotakis. Image correlation velocimetry. *Experiments in Fluids*, 19:1–15, 1995.
134. N. Fomin, V. Gamezo, and E. Lavinskaja. Analysis of simulated images of reactive flows applied to celular detonations. In *Proc. of the 16th Int. Colloquium on the Dynamic of Explosions and Reactive Systems*, page 484, Kracow, 1997. 16th ICDERS.
135. V. Gamezo, E. Lavinskaja, and N. Fomin. Image correlation analysis applied to reactive flows behind detonation waves. In B. Rinkevichus and Yu. Dubnishev, editors, *Optical Methods for Flow Studies- Proc. of the 4th Russian National Conference.*, pages 43–44, Moscow, 1997. Moscow Power Institute Press.
136. C.S. Vikram and K. Vedam. Measurement of subspeckle-size changes by laser-speckle photography. *Optics Letters*, 4(12):406–407, 1979.
137. C. Gray. *VidPIV for Microsoft Windows. Analysis Tools for Particle Image Velocimetry. Version 1.1 Demo.* Optical Flow Systems, 1994.
138. S. Debrus and C.P. Grover. Correlation of light beams scattered at different angles by a ground glass. *Optics Communications*, 3(5):340–342, 1971.
139. D.W. Li and F.P. Chiang. Decorrelation functions in laser speckle photography. *Journal of the Optical Society of America*, 3A(7):1023–1031, 1986.
140. R. Meynart. Instantaneous velocity field measurements in unsteady gas flow by speckle velocimetry. *Applied Optics*, 22:535–540, 1983.
141. R. Meynart. Non-Gaussian statistics of speckle noise of Young's fringes in speckle velocimetry. *Applied Optics*, 24(10):1448–1453, 1985.
142. C. Joenathan and R.S. Sirohi. Elimination of error in speckle photography. *Applied Optics*, 25(11):1791–1794, 1986.

143. M.S. Sivasubramanian, R. Cole, and P.C. Sukanek. Optical temperature gradient measurements using speckle photography. *International Journal of Heat and Mass Transfer*, 27(5):773–780, 1984.
144. U. Werneking, W. Merzkirch, and N.A. Fomin. Measurement of light deflection in a turbulent density field. *Experiments in Fluids*, 3:206–208, 1985.
145. U. Werneking and W. Merzkirch. Speckle photography of spatially extended refractive-index field. *Applied Optics*, 26:31–32, 1987.
146. Z.C. Liu, W. Merzkirch, and K. Oberste-Lehn. Optical tomography applied to speckle photographic measurement of axymetric flows with variable density. *Experiments in Fluids*, 7:157–163, 1989.
147. D. Kastell, K.D. Kihm, and L.S. Fletcher. Study of laminar thermal boundary layers occurring around the leading edge of a vertical isothermal wall using a specklegram technique. *Experiments in Fluids*, 13:249–256, 1992.
148. K.D. Kihm, J.H. Kim, and L.S. Fletcher. Investigation of natural convection heat transfer in converging channel flows using a specklegram technique. *Journal of Heat Transfer*, 115:140–148, 1993.
149. K.D. Kihm and S.K.R. Cheeti. Study of thermal flows from two-dimensional, upward-facing isothermal surfaces using a laser speckle photography technique. *Experiments in fluids*, 17:246–252, 1994.
150. K.D. Kihm. Laser speckle photography technique applied for heat and mass transfer problems. *Advances in Heat Transfer*, 30:255–311, 1997.
151. I. Lira. Measurement of an axisymmetric temperature field by speckle photography methods. *Experiments in Fluids*, 20:100–105, 1995.
152. G.N. Blinkov, N.A. Fomin, and D.E. Vitkin. Gasdynamic flow tomography using multidirectional speckle photography. In G.G. Levin, editor, *Analytical methods forOptical Tomography. SPIE Proceedings*, Volume 1843, pages 141–149, SPIE Press, 1992.
153. F.P. Chiang. A new three-dimensional strain analysis technique by scattered light speckle interferometry. In E.R.Robertson, editor, *The Engineering Uses of Coherent Optics: Proc. of the Conference Heldat the University of Strathclyde*, pages 249–262, Cambridge, 1976. Cambridge University Press.
154. D.B. Barker and M.E. Fourney. Displacement measurements in the interior of 3-d bodies using scattered light speckle pattern. *Journal of Experimental Mechanics*, 16(6):209–214, 1976.
155. D.B. Barker and M.E. Fourney. Measuring fluid velocities with speckle pattern. *Optical Letters*, 1:135–137, 1977.
156. T.D. Dudderar and P.G. Simpkins. Laser speckle photography in a fluid medium. *Nature*, 270(5632):45–47, 1977.
157. R. Grousson and S. Mallick. Study of the distribution of velocity in a fluid by speckle photography. *Proc. of SPIE*, Volume 136, pages 266–269, 1977.
158. R. Grousson and S. Mallick. Study of flow patterns in a fluid by scattered laser light. *Applied Optics*, 16:2334–2336, 1977.
159. J.P. Lallement, R. Desailly, and C. Froehly. Mésure de vitesse dans un liquide par diffusion cohérente. *Acta Astronautica*, 4:343–356, 1977.
160. T.D. Dudderar and P.G. Simpkins. The development of scattered light metrology. *Journal of Optical Engineering*, 21(3):396–399, 1982.
161. J.A. Leendertz. Interferometric displacement measurement on scattering surfaces utilizing speckle effect. *Journal of Physics E: Scientific Instruments*, 3:214–218, 1970.
162. K.A. Stetson. New design for laser image-speckle interferometer. *Optics and Laser Technology*, 2:179, 1970.

163. J.N. Butters and J.A. Leendertz. A double exposure technique for speckle pattern interferometry. *Journal of Physics E: Scientific Instruments*, 4:277–279, 1971.
164. P.K. Rastogi. Techniques of displacement and deformation measurements in speckle metrology. In R.S. Sirohi, editor, *Speckle Metrology*, pages 41–98. Marcel Dekker, Inc., New York, 1993.
165. R.S. Sirohi. Speckle methods in experimental mechanics. In R.S.Sirohi, editor, *Speckle Metrology*, pages 99–155. Marcel Dekker, Inc., New York, 1993.
166. Y.Y. Hung and J.D. Hovanesian. Full-field surface-strain and displacement analysis of three-dimensional objects by speckle interferometry. *Experimental Mechanics*, 12:454–460, 1972.
167. K.A. Stetson. Analysis of double exposure speckle photography with two-beam illumination. *Journal ot the Optical Society of America*, 64:857–861, 1974.
168. E. Archbold and A.E. Ennos. Application of holography and speckle photography to the measurement of displacement and strain. *Journal of Strain Analysis*, 9:10–17, 1974.
169. K. Creath. Phase-measurement interferometry technique. *Progress in Optics*, 26:349–393, 1988.
170. K. Creath. Phase-shifting speckle interferometry. *Applied Optics*, 24:3053–3058, 1985.
171. S. Nakadate. High precision retardation measurement using phase detection. *Journal of Applied Optics*, 29:242–246, 1990.
172. I. Grant and A. Liu. Accuracy consideration in incoherent analysis of piv images. *Applied Optics*, 28:4508–4510, 1989.
173. T. Rösgen, K. Wozniak, and G. Wozniak. Image processing for laser speckle velocimetry using 2-d fast fourier transform. *Applied Optics*, 29:5298–5302, 1990.
174. N. Deng and I. Yamaguchi. Automatic analysis of speckle photography with extended range and improved accuracy. *Applied Optics*, 29:296–303, 1990.
175. K.D. Hinsch. Particle image velocimetry. In R.S. Sirohi, editor, *Speckle Metrology*, pages 235–324. Marcel Dekker, Inc., New York, 1993.
176. R. Erbeck. Fast image processing with a microcomputer applied to speckle photography. *Applied Optics*, 24:3838–3841, 1985.
177. G.H. Kaufmann. Numerical processing of speckle photography data by Fourier transform. *Applied Optics*, 20:4277–4280, 1981.
178. C.S. Vikram and K. Vedam. Speckle photography of lateral sinusoidal vibrations: error due to varying halo intensity. *Applied Optics*, 20:3388–3391, 1981.
179. C.S. Vikram. Error in speckle photography of lateral sinusoidal vibrations: generalized square imaging aperture. *Applied Optics*, 21:1710–1713, 1982.
180. C.S. Vikram. Simple approach to process speckle photography data. *Optics Letters*, 7:374–375, 1982.
181. C.S. Vikram. Analysis of Young's fringes in speckle photography: generalized square imaging aperture. *Applied Physics*, 31B:221, 1983.
182. C.S. Vikram. Interpretation of Young's fringes in speckle photography for lateral vibrational analysis: imaging with circular aperture. *Optik*, 65:263–267, 1983.
183. C.S. Vikram and K. Vedam. Processing of speckle photograph data: circular imaging aperture. *Applied Optics*, 22:653–654, 1983.
184. C.S. Vikram and K. Vedam. Selective counting path of Young's fringes in speckle photography for eliminating diffraction halo effects. *Applied Optics*, 22:2242–2243, 1983.
185. R. Meynart. Diffraction halo in speckle photography. *Applied Optics*, 23:2235–2236, 1984.

186. J. Georgieva. Diffraction halo effect in speckle photography. *Applied Optics*, 25:3970–3971, 1986.
187. J. Georgieva. Removing the diffraction halo effect in speckle photography. *Applied Optics*, 28:21–22, 1989.
188. K. Hinsch. Fringe position in double exposure speckle photography. *Applied Optics*, 28:5298–5304, 1989.
189. C.S. Vikram. *Selected Papers on Holographic Particle Diagnostics*, Volume MS21 of *SPIE Milestone Series*. SPIE Press, 1990.
190. C.J.D. Pickering and N.A. Halliwell. Particle image velocimetry: fringe visibility and pedestal removal. *Applied Optics*, 24:2474–2476, 1985.
191. V. Shapiro, D. Goutev, and V. Kavardjikov. Automatic Young's fringe analysis in the source image plane. *Optics and Lasers in Engineering*, 18:307–324, 1993.
192. V.A. Shapiro, V.I. Kavardjikov, and S.A. Atanassov. Approach to automatic analysis of Young's fringes in speckle photography. *Applied Optics*, 32:4378–4387, 1993.
193. F.P. Chiang and D.W. Li. Diffraction halo functions of coherent and incoherent random speckle patterns. *Applied Optics*, 24:2166–2177, 1985.
194. Y.C. Cho. Digital image velocimetry. *Applied Optics*, 28:740–748, 1989.
195. J.M. Huntley. An image processing system for the analysis of speckle photographs. *Journal of Physics E: Scientific Instruments*, 19:43–49, 1986.
196. J. Gu and F. Chen. Fourier transformation, phase-iteration, and least-square fit image processing for Young's fringe pattern. *Applied Optics*, 35:232–239, 1996.
197. G.H. Kaufmann. Automatic fringe analysis procedures in speckle metrology. In R.S.Sirohi, editor, *Speckle Metrology*, chapter 9, pages 427–472. Marcel Dekker, Inc., New York, 1993.
198. J.M. Huntley. Speckle photography fringe analysis: assessment of current algorithms. *Applied Optics*, 28:4316–4322, 1989.
199. D.W. Robinson. Automatic fringe analysis with a computer image processing system. *Applied Optics*, 22:2169–2176, 1983.
200. J.M. Huntley, H.T. Goldrein, and L.R. Benckert. Parallel processing system for rapid analysis of speckle photography and particle image velocimetry data. *Applied Optics*, 32:3152–3155, 1993.
201. J.M. Huntley. Speckle photography fringe analysis by the Walsh transform. *Applied Optics*, 25:382–386, 1986.
202. J.L. Walsh. A closed set of orthogonal functions. *Ann. J. Math.*, 55:5, 1923.
203. K.G. Beauchamp. *Walsh Functions and their Applications*. Academic Press, New York, 1975.
204. D.F. Elliot and K.R. Rao. *Fast Transforms*. Academic Press, New York, 1982.
205. M.O. Petersen. Digital speckle pattern shiftinginterferometry: limitations and prospects. *Applied Optics*, 30:2730–2738, 1991.
206. J. Gu, Y.Y. Hung, and F. Chen. Iteration algorithm for computer-aided speckle interferometry. *Applied Optics*, 33:5308–5317, 1994.
207. M. Takeda, H. Ina, and S. Kabayashi. Fourier transform method of fringe-pattern analysis for computer-based tomography and interferometry. *Journal of the Optical Society of America*, 72:156–162, 1981.
208. G.M. Carlomagno and A. Rapillo. Interferogram analysis by means of photodiode array. In *Optics in Engineering Measurements. Proc. SPIE*, Volume 599, pages 160–165. SPIE Press, 1985.
209. J.C. Hunter and M.W. Collins. Three-dimensional refractive index field reconstruction from holographic interferograms. *International Journal of Optoelectronics*, 4(2):95–132, 1989.

210. G.E. Maddux, S.L. Moorman, and R. Corwin. A programmable data-retrieval system for in-plane displacement from speckle photographs. Technical report, US Air Force Technical Report AFFDL-TM-78-109-FBE, 1978.
211. G.H. Kaufmann, A.E. Ennos, B. Gale, and D.J. Pugh. An electro-optical readout system for analysis of speckle photographs. *Journal of Physics E: Scientific Instruments*, 13:579–584, 1980.
212. B. Ineichen, P. Eglin, and R. Dăndliker. Hybrid optical and electronic image processing for strain measurements by speckle photography. *Applied Optics*, 19:2191–2195, 1980.
213. R. Meynart. Digital image processing for speckle flow velocimetry. *Reviw of Scientific Instruments*, 53:110–111, 1982.
214. R. Erbeck. *Die Anwendung der Speckle-Photographie zur statistischen Analyse turbulenter Dichtefelder*. PhD dissertation, Essen University, 1986.
215. S. Toyooka, Y. Iwaasa, M. Kawahashi, K. Hosoi, and M. Suzuki. Automatic processing of Young's fringes in speckle photography. *Optics and Lasers in Engineering*, 6:203–212, 1985.
216. K. Hinsch and W. Arnold. Spatially multiplexed optical correlation for displacement mapping in speckle metrology. In *Proc. SPIE*, Volume 1136, pages 327–334, 1989.
217. P. Buchhave. Optical processing of PIV images. *Journal of Flow Visualization and Image Processing*, 2:149–160, 1995.
218. C.S. Yao and R.J. Adrian. Orthogonal compression and 1-d analysis technique for measurement of 2-d particle displacements in pulsed laser velocimetry. *Applied Optics*, 23:1687–1689, 1984.
219. S.H. Collicott and L. Hesselink. Anamorphic optical processing of multiple-exposure speckle photographs. *Optics Letters*, 11:410–412, 1986.
220. S.H. Collicott and L. Hesselink. Analysis and design of an anamorphic optical processor for speckle metrology and velocimetry. *Applied Optics*, 31:1646–1659, 1992.
221. W. Arnold and K.D. Hinsch. Parallel optical evaluation of double exposure records in optical metrology. *Applied Optics*, 28(4):726–729, 1989.
222. S.H. Collicott and L. Hesselink. Real-time photorefractive recording and optical processing for speckle velocimetry. *Optics Letters*, 13:348, 1988.
223. J.M. Coupland and N.A. Halliwell. Particle image velocimetry: rapid transparency analysis using optical correlation. *Applied Optics*, 27:1919–1921, 1988.
224. P. Buchhave and M.L. Jakobsen. Optical correlator for piv image processing. In *Laser Anemometry, Advances and Applications, Proc. of the 3rd Int. Conference*, pages 609–616. Springer Verlag, 1990.
225. B. Bates and P.C. Miller. Liquid crystal television in speckle metrology. *Applied Optics*, 27:2816–2817, 1988.
226. J.P. Sharpe and K.M. Johnson. Particle image velocimetry fringe processing using an optically addressed spatial light modulator. *Applied Optics*, 31:7399–7402, 1992.
227. M.J. Jakobsen, W.J. Hossak, C.A. Greated, and W.J. Easson. PIV analysis using an optically addressed spatial light modulator. *Optics and Lasers in Engineering*, 19:253–260, 1993.
228. D. Gunningham, J. Sharpe, and K.M. Johnson. Application of an optically addressed spatial light modulator to real time speckle photography. *Optics Communications*, 101:311–316, 1993.
229. Z.Q. Mao, N.A. Halliwell, and J.M. Coupland. Particle image velocimetry: High speed transparency scanning and correlation peak location in optically processing system. *Applied Optics*, 32:5089–5091, 1993.

230. M.J. Jakobsen, W.J. Hossak, and C.A. Greated. PIV analysis using an optically addressed spatial light modulator: Effect of nonlinear transfer function. *Applied Optics*, 34(11):1757–1762, 1995.
231. L. Celaya, J.M. Jonathan, and S. Mallick. Velocity contours by speckle photography. *Optics Communications*, 18:496–498, 1976.
232. U. Gärtner, U. Wernekinck, and W. Merzkirch. Velocity measurements in the field of an internal gravity wave by means of speckle photography. *Experiments in Fluids*, 4:283–287, 1986.
233. W. Merzkirch and H. Wintrich. Flow visualization by spatial filtering. In R. Reznichek, editor, *Flow Visualization V*, pages 183–187, Washington, DC, 1990. Hemisphere Publ. Co.
234. V. Palero, N. Andres, M.P. Arroyo, and M. Quintanilla. Fast quantitative processing of particle image velocimetry photographs by a whole-field filtering technique. *Experiments in Fluids*, 19:417–425, 1995.
235. W. Merzkirch. Density-sensitive flow visualization. In R.J. Emrich, editor, *Fluid Dynamics*, Volume 18A of *Methods of Experimental Physics*, pages 345–403. Academic Press, New York, 1981.
236. D. Bershader. Refractive behavior of gases. In *Modern Optical Methods*, page 65. Plenum Press, New York, 1971.
237. D. Bershader and S.G. Prakash. Progress on improved flow visualization by resonance refractivity. In *Combustion Measurements*, pages 245–252. Academic Press, New York, 1976.
238. E. Kügler and D. Bershader. Recent high-resolution resonant refractivity studies of a sodium-seeded flame. *Experiments in Fluids*, 1:51–55, 1983.
239. H. Fiedler, K. Nottmeyer, P.P. Wegener, and S. Raghu. Schlieren photography of water flow. *Experiments in Fluids*, 3:145–151, 1985.
240. H.M. Dobbins and E.R. Peck. Change of refractive index of water as a function of temperature. *Journal ot the Optical Society of America*, 63:318–320, 1973.
241. F. Peters. Schlieren interferometry applied to a gravity wave in a density-stratified liquid. *Experiments in Fluids*, 3:261–269, 1985.
242. Yu.A. Kravtsov and Yu.I. Orlov. *Geometrical Optics of Inhomogeneous Media*, Volume 6 of *Wave Phenomena*. Springer Verlag, Berlin, 1994.
243. F.J. Weyl. Analysis of optical methods. In *Physical Measurements in Gas Dynamics and Combustion*, pages 3–25. Princeton University Press, Princeton, 1954.
244. A. Sharma, D.V. Kumar, and A.K. Ghatak. Tracing rays through graded-index media: a new method. *Applied Optics*, 21(6):984–987, 1982.
245. A. Sharma. Computing optical path lengh in gradient-index media: a fast and accurate method. *Applied Optics*, 24(24):4367–4370, 1986.
246. S. Dorić. Ray tracing through gradient-index media: recent improvements. *Applied Optics*, 29(28):4026–4029, 1990.
247. E. Lavinskaja, E.F. Nogotov, and N.A. Fomin. Statistical analysis of turbulence by speckle photography. *Reports of the Belarus Academy of Sciences*, 41(4):53–57, 1997.
248. N.A. Fomin, and D.E. Vitkin. Turbulence parameters measurements by speckle technique (in Russian). *Journal of Physical Engineering*, 69(5):805–810, 1996.
249. G.N. Blinkov, D.E. Vitkin, V.E. Fertman, and N.A. Fomin. Speckle-interferometry of the thermal gravity convection in a norrow gap (in Russian). In *Proc. of the Minsk International Forum on Heat and Mass Transfer, Selected papers, Sections 1-2, part 1*, pages 11–23, Minsk, Belarus, 1989. The Heat and Mass Transfer Institute.

250. U. Wernekinck, W. Merzkirch, and N. Fomin. Speckle photography of gasdynamic fields in turbulent reactive flows (in Russian). In O. Martynenko, editor, *Proc. Seventh All-Union Heat and Mass Transfer Conference (USSR)*, Volume 3, pages 45–53. Heat and Mass Transfer Institute Press, Minsk, 1984.
251. N.A. Fomin, U. Wernekinck, and W. Merzkirch. Speckle photography of a turbulent density field. In M. Píchal, editor, *Proc. UTIAM-Symposium on the Optical Methods in the Dynamics of Fluids and Solids*, pages 159–165. Springer Verlag, Berlin, 1985.
252. U. Wernekinck and W. Merzkirch. Measurement of natural convection by speckle photography. In C.L. Tien, V.P. Carey, and J.K. Ferrell, editors, *Heat Transfer 1986*, pages 531–535. Hemisphere Publ. Co., Washington DC, 1986.
253. N.A. Fomin. Optical diagnostic activities at the Belarus Heat and Mass Transfer Institute. In *Optical Technology in Fluid, Thermal, and Combustion Flow III. SPIE Proc.*, Volume 3172-50, SPIE Press, Bellingham, 1997.
254. K.D. Kihm. Image blurring of test section boundary in a laser specklegram technique measuring temperature gradients of a compressible medium. *Applied Optics*, 31:5907–5910, 1992.
255. T.E. Walsh and K.D. Kihm. Laser speckle photography technique for measuring local and global heat transfer correlation coefficients. *Journal of Flow Visualiztion and Image Processing*, 1(4):359–370, 1993.
256. J.Z. Shu and J.Y. Li. Speckle photography applied to the density field of a flame. *Experiments in Fluids*, 5:422–424, 1987.
257. J.Z. Shu and J.Y. Li. A laser schlieren speckle interferometry system for measurement of phase objects. *Journal of Flow Visualization and Image Processing*, 1:63–68, 1993.
258. R.I. Soloukhin. Measurement of the rate of oxygen recombination in shock waves. *Combustion, Explosions, and Shock Waves*, 3:402–411, 1967.
259. O. Achasov, N.A. Fomin, O. Penyazkov, and F. Fisson. Interferometric study of combustion in a spark ignition engine. In *Annual Proc. of the Heat and Mass Transfer Institute*, pages 65–69, Minsk, 1990. Heat and Mass Transfer Institute Press.
260. O.V. Achasov, G.N. Blinkov, N.A. Fomin, O.G. Penyazkov, and D.E. Vitkin. Speckle tomography of unsteady gasdynamic objects. In *Dynamics of Deflagrations and Reactive Systems: Heterogeneous Combustion. Proc. of the Twelfth Int. Coll. on Dynamics of Explosions and Reactive System, Ann Arbor, 1989*, Volume 132 of *Progress in Astronautics and Aeronautics*, pages 553–562, Washington, DC, 1991. AIAA-Press.
261. D.D. Verhoeven and P.V. Farrell. Speckle interferometry in transparent media. *Applied Optics*, 25(6):903–906, 1986.
262. S. Urgela and P. Kubošek. Measurement of a 2D temperature field by electronic speckle pattern interferometry. *Journal of Flow Visualization and Image Processing*, 1(2):85–89, 1993.
263. Y. Song, R. Kulenovich, M. Groll, and Z. Guo. Effects of speckle displacement on speckle interferometry for measurement of phase object. *Optics Communications*, 139:24–30, 1997.
264. Y.Z. Song, R. Kulenovic, M. Groll, and Z.Y. Guo. Digital schearing speckle interferometry applied to optical diagnostics in flow. In *Optical Technology in Fluid, Thermal, and Combustion Flow III. SPIE Proc.*, Volume 3172-23, Bellingham, 1997. SPIE Press.
265. Y.Z. Song, R. Kulenovic, Z.Y. Guo, and M. Groll. Digital speckle pattern interferometry using polarization phase shift for measurement of temperature field in thermal flow. In *Optical Technology in Fluid, Thermal, and Combustion Flow III. SPIE Proc.*, Volume 3172-46, Bellingham, 1997. SPIE Press.

266. C. Shakher, A.K. Nirala, J. Pramila, and S.K. Verma. Use of speckle technique for temperature measurement in gaseous flame. *J. Optics (Paris)*, 23:35–39, 1992.

267. S. Shakher and A.K. Nirala. A comparative study of speckle shearing interferometry and speckle photography for measurement of temperature of an axisymmetrc flame. *Optik*, 97:43–46, 1994.

268. S. Shakher and A.K. Nirala. Measurement of temperature using speckle shearing interferometry. *Applied Optics*, 33:2125–2127, 1994.

269. S. Shakher and A.K. Nirala. Temperature profile measurement of axisymmetric gaseous flames using speckle photography, speckle shearing interferometry, and Talbot interferometry. *Optical Engineering*, 33(6):1983–1988, 1994.

270. T.S. Durrani and C.A. Greated. *Laser Systems in Flow Measurements*. Plenum Press, New York, 1977.

271. L.E. Drain. *The Laser Doppler Technique*. John Willey & Sons, New York, 1980.

272. F. Durst, A. Melling, and J.H. Whitelaw. *Principles and Practice of Laser Doppler Anemometry*. Academic Press Inc., New York, second edition, 1987.

273. C.J. Chen and R.J. Emrich. Investigation of schock tube boundary layer by a tracer method. *Physics of Fluids*, 6:1–9, 1963.

274. R.J. Emrich. Flow field measurement by tracer photography. *Experiments in Fluids*, 1:179–184, 1983.

275. R.J. Perkins and J.C.R. Hunt. Particle tracking in turbulent flows. In *Advances in Turbulence*, Volume 2, pages 286–291. Springer Verlag., Berlin, 1989.

276. B. Khalighi. Study of the intake tumble motion by flow visualization and particle tracking velocimetry. *Experiments in Fluids*, 10:230–236, 1991.

277. R.J. Adrian. Image shifting technique to resolve directional ambiguity in double-pulsed velocimetry. *Applied Optics*, 25(21):3855–3858, 1986.

278. R.J. Adrian. Particle-imaging techniques for experimental fluid mechanics. *Annual Reviw of Fluid Mechanics*, 23:261–304, 1991.

279. I. Grant. Particle image velocimetry: a review. In *Proc. of Inst. Mech. Engrs., Part C*, Volume 211, pages 55–76, London, 1997. IMech Press.

280. M.P. Wernet. PIV for turbomachinery applications. Technical memorandum 107525, NASA, Lewis Research Center, Clevelend, Ohio, 1997.

281. K.D. Hinsch. Three-dimensional particle velocimetry. *Meas. Sci. Technol.*, 6:742–753, 1995.

282. L. Lourenço and A. Krothapalli. Particle image displacement velocimetry measurements of a three-dimensional jet. *Physics of Fluids*, 31:1835–1837, 1988.

283. Z.C. Liu, C.C. Landreth, R.I. Adrian, and T.J. Hanratty. High resolution measurements of turbulent structure in a channel with particle image velocimetry. *Experiments in Fluids*, 10:301–312, 1991.

284. C. Gray, D. Skyner, and C.A. Greated. The measurement of breaking waves using particle image velocimetry. In R.J.Adrian, editor, *Proc. ICALEO'88*, Volume LIA 67, pages 166–177, Toledo OH, 1988. Laser Institute of America.

285. C. Gray, C.A. Greated, D.R. McCluskey, and W.J. Easson. An analysis of the scanning beam piv illumination system. *Engineering Optics*, 4:461–468, 1991.

286. M. Kawahashi and K. Hosoi. Beam-sweep laser speckle velocimetry. *Experiments in Fluids*, 8:109–111, 1989.

287. R.J. Adrian. Scattering particle characteristics and their effect on pulsed laser measurements of fluid flow: speckle velocimetry vs. particle image velocimetry. *Applied Optics*, 23:1690–1691, 1984.

288. C.C. Landreth and R.J. Adrian. Electrooptical image shifting for particle image velocimetry. *Applied Optics*, 27(20):4216–4220, 1988.

289. J.M. Coupland, C.J.D. Pickering, and N.A. Halliwell. Particle image velocimetry: theory of the directional ambiguity removal using holographic image separation. *Applied Optics*, 26(9):1576–1578, 1987.
290. J.M. Coupland, C.J.D. Pickering, and N.A. Halliwell. Particle image velocimetry: the ambiguity problem. *Journal of Optical Engineering*, 27(3):193–196, 1988.
291. B. Ruck. A new laser-optical method for visualization and real-time vectorization of flow fields. In *Optical Methods and Data Processing in Heat and Fluid Flow*, pages 447–456, London, 1996. IMechE Press.
292. M.P. Wernet and R.V. Edwards. New space domain processing technique for pulsed laser velocimetry. *Applied Optics*, 29:3399–13417, 1990.
293. E.J. Klein. Liquid crystal in aerodynamic testing. *Astronautics & Aeronautics* , 6:70–83, 1968.
294. T.E. Cooper and R.J. Field. Liquid crystal thermography and its application to the study of convective heat transfer. *Journal of Heat Transfer*, 97:442–450, 1975.
295. T.R. Ogden and E.W. Hendrics. Liquid crystal thermography in water tunnels. *Experiments in Fluids*, 2:65–66, 1984.
296. S. Chandrasekhar. *Liquid Crystals.* Cambridge Univ. Press, London, 1977.
297. G.W. Gray. *Thermotropic Liquid Crystals.* John Willey & Sons, New York, 1987.
298. J.A. Szymczyk, J. Siekmann, Ch-H. Chun, and K. Wozniak. Liquid crystal tracers as a method for thermocapillary flow diagnostics. *Archivum of Mechanics*, 41(2), 1989.
299. I. Kimura, T. Takamori, M. Ozawa, N. Takenaka, and T. Sakaguchi. Simultaneous measurement of flow and temperature fields based on color image information. In R. Řezniček, editor, *Flow Visualization V*, pages 29–41, New York, 1990. Hemisphere Publishing Corporation.
300. D. Daribi and M. Gharib. Digital particle image thermometry: Method and implementation. *Experiments in Fluids*, 11:77–86, 1991.
301. M. Ozawa, U. Müller, I. Kimura, and T. Takamori. Flow and temperature measurement of natural convection in a Hele–Shaw cell using a thermo-sensitive liquid-crystal tracer. *Experiments in Fluids*, 12:213–222, 1992.
302. G. Wozniak and K. Wozniak. Buoyancy and thermocapillary flow analysis by the combined use of liquid crystals and PIV. *Experiments in Fluids*, 17:141–146, 1994.
303. J.O. Hinze. *Turbulence.* McGraw-Hill Book Inc., New York, second edition, 1975.
304. I. Grant, E. Owens, Y.-Y. Yan, and X. Shen. Particle image velocimetry measurements of the separated flow behind a rearward facing step. *Experiments in Fluids*, 12:238–244, 1992.
305. W. Arnold, K.D. Hinch, and D. Mach. Turbulence level measurements by speckle velocimetry. *Applied Optics*, 25:330–331, 1986.
306. R.J. Adrian. Statistical properties of particle image velocimetry measurements in turbulent flows. In *Laser Anemometry in Fluid Mechanics*, Volume 3, pages 115–129, Berlin, 1988. Springer Verlag.
307. K. Hinsch, W. Schipper, and D. Mach. Fringe visibility in speckle velocimetry and the analysis of random flow components. *Applied Optics*, 23:4460–4462, 1984.
308. K. Hinsch, W. Arnold, and W. Platen. Flow field analysis by large area interrogation in particle image velocimetry. *Optics and Lasers in Engineering*, 9:229–243, 1988.

309. K. Hinsch, W. Arnold, and W. Platen. Turbulence measurements by particle imaging velocimetry. In W.H. Stevenson, editor, *Proc. ICALEO'87*, Volume LIA 63, pages 127–134, Toledo, OH, 1988. Laser Institute of America.
310. I. Grant, E.H. Owens, and G.H. Smith. Particle image velocimetry fringe visibility as a means of turbulence estimation. *Applied Optics*, 28:2982–2984, 1989.
311. P. Jacquot and P.K. Rastogi. Influence of out-of-plane deformation and its elimination in white-light speckle photography. *Optics and Lasers in Engineering*, 2:33–55, 1981.
312. R.G. Racca and J.M. Devey. A method for automatic particle tracking in a three-dimensional flow field. *Experiments in Fluids*, 6:25–32, 1988.
313. A.A. Adamczyk and L. Rimai. 2-dimensional particle tracking velocimetry (PTV): Technique and image processing algorithms. *Experiments in Fluids*, 6:373–380, 1988.
314. V. Gauthier and M.L. Riethmuller. *Application of PIDV to Complex Flows: Measurement of the Third Component.* VKI Lectures Series on Particle Image Displacement Velocimetry. Brussels, 1988.
315. M.P. Arroyo and C. Greated. Stereoscopic particle image velocimetry. *Measurement Science and Technology*, 2:1181–1186, 1991.
316. I. Grant, Y. Zhao, Y. Tan, and J. N. Stewart. Three component flow mapping: Experiences in stereoscopic PIV and holographic velocimetry. *Laser Anemometry*, 1:365–371, 1991.
317. A.K. Prasad and R.J. Adrian. Stereoscopic particle image velocimetry applied to liquid flows. *Experiments in Fluids*, 15:49–60, 1993.
318. V. Palero, M.L. Jakobsen, M.P. Arroyo, and M. Quintanilla. Stereoscopic particle image velocimetry for size and velocity measurements in multi-phase flows. In *6th International Conferences on Industrial metrology CIMI'95*, pages 462–470, Zaragoza University, 1995.
319. I. Grant, S. Fu, X. Pan, and X. Wang. The application of an on-line, stereoscopic, PIV system to 3-component velocity measurements. *Experiments in Fluids*, 19:214–221, 1995.
320. A. Cenedese and A. Paglialunga. A new technique for the determination of the third velocity component with PIV. *Experiments in Fluids*, 8:228–230, 1989.
321. J.D. Trolinger. Particle field holography. *Optical Engineering*, 14(5):383–392, 1975.
322. C.S. Moratis and P. Buchhave. Three dimensional instantaneous velocity measurements by optical processing of holographic records. In *Laser Anemometry Advances and Applications, SPIE Proc.*, Volume 2052, Bellingham, 1993. SPIE Press.
323. C. Gray and C.A. Greated. Practical application of particle field holography for flow velocimetry and strain measurement. In R. Royles, editor, *Advances in Engineering Measurements. Proc. of the Int. Conf.*, pages 54–57, Edinburgh, 1994. The University of Edinburgh.
324. H. Meng and F. Hussain. In-line recording and off-axis viewing technique for holographic particle velocimetry. *Applied Optics*, 34(11):1827–1840, 1995.
325. J. Zhang, B. Tao, and J. Katz. Turbulent flow measurement in a square duct with hybrid holographic PIV. *Experiments in Fluids*, 23:373–381, 1997.
326. K.D. Hinsch. The many dimensions of optical flow diagnostics. *Laser Anemometry Advances and Applications. Proc. of SPIE*, Volume 2052, pages 63–78, 1993, SPIE Press.
327. L.P. Bernal and J. Schrer. *HPIV measurements in vortical flows*, Volume 148, pages 43–50. Fluids Engineering Division (American Society of Mechanical Engineers, 1993.

328. S. Simmons, H. Meng, F. Hussain, and D. Liu. Advances of holographic particle velocimetry. In *Optical Diagnostics in Fluid and Thermal Flow*, Volume 2005, pages 306–317. SPIE Press, 1993.
329. D.H. Barnhart, R.J. Adrian, and G.C. Papen. Phase conjugate holographic system for high resolution particle image velocimetry. *Applied Optics*, 33:7159–7170, 1994.
330. H. Meng and F. Hussain. Instantaneous flow field in an unstable vortex ring measured by holographic particle velocimetry. *Physics of Fluids*, 7(1):9–11, 1995.
331. J.S. Slepicka and S.S. Cha. Three-dimensional flow diagnostics by holographic diffraction image velocimetry. In *Optical Technology in Fluid, Thermal, and Combustion Flow III. SPIE Proc.*, Volume 3172-32, Bellingham, 1997. SPIE Press.
332. N.H. Abramson. Picosecond pulses for holographic sectioning of volumes. In *Optical Technology in Fluid, Thermal, and Combustion Flow III. SPIE Proc.*, Volume 3172-17, Bellingham, 1997. SPIE Press.
333. N. Andrés, M.P. Arroyo, and M. Quintanilla. Holographic interferometry velocity measurements in a convective flow. In *European Optical Society. Annual Meetings Digest Series*, Volume 3, pages 137–138, Zaragoza, 1993.
334. N. Andrés, M.P. Arroyo, and M. Quintanilla. Three component velocity measurements in a convective flow by holographic interferometry. In *7th International Symposium on Applications of Laser Techniques to Fluid Mechanics*, pages 32.4.1–32.4.6, Lisbon, 1994.
335. B. Skarman, J. Becker, and K. Wozniak. Simultaneous 3D-PIV and temperature measurements using a new CCD - based holographic interferometer. *Flow Measurement Instruments*, 7:1–6, 1996.
336. N. Takai, T. Iwai, T. Ushizaka, and T. Asakura. Velocity measurement of the diffuse object based on time-differentiated speckle intensity fluctuations. *Optics Communications*, 30(3):287–292, 1979.
337. N. Takai, T. Iwai, and T. Asakura. Real-time velocity measurement for a diffuse object using zero-crossings of laser speckle. *Journal of the Optical Society of America*, 70(4):450–455, 1980.
338. H. Fujii, T. Asakura, K. Nohira, Y. Shintomi, and T. Ohura. Blood flow observed by time-varrying laser speckle. *Optics Letters*, 10(3):104–106, 1985.
339. Y. Aizu and T. Asakura. Bio-speckle phenomena and their application to the evaluation of blood flow. *Optics and Laser Technology*, 23(4):205–219, 1991.
340. R. Erbeck and W. Merzkirch. Speckle photographic measurements of turbulence in an air stream with fluctuating temperature. *Experiments in Fluids*, 6:89–93, 1988.
341. J. Keller and W. Merzkirch. Interaction of a normal shock wave with a compressible turbulent flow. *Experiments in Fluids*, 8:241–248, 1990.
342. K. Oberste-Lehn and W. Merzkirch. Speckle optical measurement of a turbulent scalar field with high fluctuation amplitude. *Experiments in Fluids*, 14(4):217, 1993.
343. S. Simoens and M. Ayrault. Concentration flux measurements of a scalar quantity in turbulent flows. *Experiments in Fluids*, 16:273–6281, 1994.
344. Z. Warhaft and J.L. Lumley. An experimental study of the decay of temperature fluctuations in grid-generated turbulence. *Journal of Fluid Mechanics*, 88:659–684, 1978.
345. T. Grandke. Theory and application of the laser shadow technique. part 1: Two-dimensional turbulent flow. *Experiments in Fluids*, 3:77–86, 1985.
346. T. Grandke. Theory and application of the laser shadow technique. part 2: Axisymmetric turbulent flow. *Experiments in Fluids*, 3:277–282, 1985.

347. M.M. Stanišić. *The Mathematical Theory of Turbulence.* Springer-Verlag, New-York, 1988.
348. A.V. Smol'yakov and V.M. Tkachenko. *The Measurement of Turbulent Fluctuations. An Introduction to Hot-Wire Anemometry and Related Transducers.* Springer Verlag, Berlin, 1983.
349. A.N. Kolmogorov. The local structure of turbulence in incompressible viscous fluids for very large Reynolds numbers. *USSR Academy Reports*, 30(4):299–313, 1941.
350. A.M. Obuchov. Structure of temperature field in turbulent flow. *Izvestia of the USSR Acad. of Sci., Seria Geogr. and Geophys.*, 8(1):58–69, 1949.
351. L. Hesselink and B. Sturtevant. Propagation of weak shocks through a random medium. *Journal of Fluid Mechanics*, 196:513–553, 1988.
352. N. Fomin and C. Greated. Acoustic wave tracing in turbulent flows. In *Proc. of II Intern. School-Seminar on Modern Problems of Combustion and its Applications*, pages 108–112, Minsk, 1997. Heat and Mass Transfer Institute Press.
353. S.M. Rhytov, U.A. Kravtsov, and V.I. Tatarskii. *Introduction to Statistical Radiophysics. Volume 2. Random Fields.* Nauka, Moscow, 1978 (in Russian).
354. T. Gerz and U. Schumann. Direct simulation of homogeneous turbulence and gravity waves in sheared and unsheared stratified flows. In *Turbulent Shear Flows 7*, pages 29–45. Springer Verlag, Berlin, 1991.
355. M.S. Uberoi and L.S.G. Kovasznay. Analysis of turbulent fluctuations by the shadow method. *Journal of Applied Physics*, 26(1):19–22, 1955.
356. L.S. Taylor. Analysis of turbulence by shadograph. *AIAA Journal*, 8(7):1284–1287, 1969.
357. R.E. Slattery and W.G. Clay. Measurement of turbulent transition, motion, statistics, and gross radial growth behind hypersonic objects. *The Physics of Fluids*, 5:849–855, 1962.
358. W.G. Clay, J. Herrman, and R.E. Slattery. Statistical properties of the turbulent wake behind hypervelocity spheres. *The Physics of Fluids*, 8(10):1792–1801, 1965.
359. J. Herrman, W.G. Clay, and R.E. Slattery. Gas-density fluctuations in the wakes from hypersonic spheres. *The Physics of Fluids*, 11(5):954–959, 1968.
360. M.R. Davis. Measurements in a subsonic jet using a quantitative schlieren technique. *Journal of Fluid Mechanics*, 46:631–656, 1971.
361. R. Erbeck. *Die Anwendung der Speckle-Photographie zur statistischen Analyse turbulenter Dichtefelder*, Volume Reihe 8, Nr.112. VDI-Forshungsbericht, 1986.
362. R. Erbeck and W. Merzkirch. Speckle photographic measurements of turbulence in an air stream with fluctuating temperature. In *Proc. 1st Symposium on Turbulence.* University of Missory-Rolla, 1986.
363. W. Merzkirch. Measurement techniques in aerodynamics. In *Laser Methods in Fluid Mechanics*, Von Karman Institute for Fluid Dynamics, Lectures Series 1989-05. 1989.
364. K. Oberste-Lehn. *Speckle-photographische Untersuchungen eines turbulenten passiven Skalarfeldes bei hohen Amplituen der Temperaturfluktuationen.* PhD dissertation, Essen University, 1991.
365. D. Vitkin, W. Merzkirch, and N. Fomin. Quantitative visualization of the change of turbulence structure caused by a normal shock wave. *Journal of Visualization*, Volume 1, in press, 1998.
366. S. Lee, S.K. Lele, and P. Moin. Direct numerical simulation of isotropic turbulence interacting with a weak shock wave. *Journal of Fluid Mechanics*, 251:533–562, 1993.

367. R. Hannapel and R. Friedrich. Interaction of isotropic turbulence with a normal shock wave. *Applied Scientific Research*, 51:507–512, 1993.
368. K. Mahesh, S.K. Lele, and P. Moin. The influence of entropy fluctuations on the interaction of turbulence with a shock wave. *Journal of Fluid Mechanics*, 334:353–397, 1997.
369. M.M. Weiner. Atmospheric turbulence in optical surveillance systems. *Applied Optics*, 6(11):1984 –1991, 1967.
370. J. Bashir and M.S. Uberoi. Experiment on turbulent structure and heat transfer in a two-dimensional jet. *The Physics of Fluids*, 18(4):405–410, 1975.
371. N.A. Fomin and E.F. Nogotov. Statistical analysis of turbulence by single exposure speckle photography. In *Optical Technology in Fluid, Thermal, and Combustion Flow III. SPIE Proc.*, Volume 3172-38, 1997, SPIE Press.
372. N. Fomin. Speckle frame- and streak-cinematography: an insight into the structure and development of turbulence. *Archivum Combustionis*, 15(3-4):227–239, 1995.
373. N.A. Fomin and D. Vitkin. Speckle tomography of temperature distribution in multi-jet turbulent flames. In *IMexE Conference Transaction. International Seminar on Optical Methods and Data Processing in Heat and Fluid Flows*, pages 121–127, London, 1996. Institute of Mechanical Engineers.
374. G.T. Herman. *Image Reconstruction from Projections: The Fundamentals of Computerized Tomography.* Academic Press, New York, 1980.
375. A.C. Kak and M. Slaney. *Principles of Computerized Tomographic Imaging.* IEEE-Press, New York, 1988.
376. V.V. Pickalov and N.G. Preobrazhensky. *Reconstructive Tomography in Gasdynamics and Plasma Physics.* Nauka. Sib.otd-nie, Novosibirsk, 1987.
377. V.V. Pickalov and T.C. Mel'nikova. *Tomography of Plasma (in Russian).* Nauka. Sib.otd-nie, Novosibirsk, 1995.
378. F. Natterer. *The Mathematics of Computerized Tomography.* John Wiley and Sons, Inc., New York, 1986.
379. E. Wolf. Three dimensional structure determination of semi-transparent objects from holographic data. *Optics Communications*, 1:153–156, 1969.
380. K. Iwata and R. Nagata. Calculation of three-dimensional refractive-index distribution from interferograms. *Journal of the Optical Society of America*, 60:133–135, 1970.
381. R.D. Matulka and D.J. Collins. Determination of three-dimensional density fields from holographic interferograms. *Journal of Applied Physics*, 42(3):1109–1119, 1971.
382. D.W. Sweeney and C.M. Vest. Reconstruction of three-dimensional refractive fields from multidirectional interferometric data. *Applied Optics*, 12:2649–2664, 1973.
383. C.M. Vest. Formation of images from projections: Radon and Abel transforms. *Journal of the Optical Society of America*, 64:1215–1218, 1974.
384. F.P. Chen and R. Goulard. Retrieval of arbitrary concentration and temperature fields by multiangular scanning techniques. *Journal of Quantitative Spectroscopy amd Radiative Transfer*, 16(10):819–827, 1976.
385. R. Goulard and P.J. Emmerman. Absorption diagnostics. In *AIAA Papers*, number AIAA Paper 79-0085, New Orleans, LA, 1979. 17th Aerospace Sciences Meeting.
386. P.J. Emmerman, R. Goulard, R.J. Santoro, and H.G. Semerjian. Multiangular absorption diagnostics of a turbulent argon-methane jet. *Journal of Energy*, 4(2):70–77, 1980.

387. R.J. Santotro and H.H. Semirjian. Optical tomography for flow field diagnostics. *International Journal of Heat and Mass Transfer*, 24(7):1139–1150, 1981.
388. K.E. Bennet, G.W. Faris, and R.L. Byer. Experimental optical fan beam tomography. *Applied Optics*, 23:2678–2685, 1984.
389. M. Hino, T. Aono, M. Nakajima, and S. Yuta. Light emission computed tomography system for plasma diagnostics. *Applied Optics*, 26:4742–4746, 1987.
390. D.W. Sweeney and C.M. Vest. Measurement of three-dimensional temperature fields about heated surfaces by holographic interferometry. *International Journal of Heat and Mass Transfer*, 17:1443–1454, 1974.
391. D.W. Watt and C.M. Vest. Turbulent flow visualization by interferometric integral imaging and computed tomography. *Experiments in Fluids*, 8:301–311, 1990.
392. D.D. Verhoeven. Application of computed tomography and holographic interferometry to the study of transparent media. Research Report 38501, Institute Franç ais du Pétrole, BP311, 92500 Rueil-Malmaison, France, 1990.
393. R.I. Soloukhin and N.A. Fomin. Speckle-photography of density field in gas reactive flows. In V.K. Baev, editor, *Gas Flame Structure*, pages 50–58, Novosibirsk, 1986. Institute of Theoretical and Applied Mechanics.
394. G.N. Blinkov, D.E. Vitkin, R.I. Soloukhin, and N.A. Fomin. Speckle photography of local temperature fields in axisymmetrical heat and mass transfer processes. *Doklady AN BSSR*, 31(10):985–898, 1987.
395. G.N. Blinkov, N.A. Fomin, and R.I. Soloukhin. Multidirectional speckle photography of density gradients in a flame. In *Dynamics of Reactive Systems. Part 1: Flames*, Volume 133 of *Progress in Astronautics and Aeronautics*, pages 403–416. AIAA, Inc., Washington, 1988.
396. B. Lu, X. Yang, H. Abendroth, H. Eggers, and E. Ziolkowski. Measurement of a three-dimensional temperature field applying ESPI and CT techniques. *Optics Communications*, 69:6–10, 1988.
397. G.N. Blinkov N.A. Fomin, M.N. Rolin, R.I. Soloukhin, D.E. Vitkin, and N.L. Yadrevskaya. Speckle tomography of a gas flame. *Experiments in Fluids*, 8:72–76, 1989.
398. C.M. Vest. Tomography for properties of materials that bend rays: a tutorial. *Applied Optics*, 24:4089–4094, 1985.
399. S.S. Cha and C.M. Vest. Tomographic reconstruction of strongly refracting fields and its application to interferometric measurement of boundary layers. *Applied Optics*, 20:2787–2794, 1981.
400. I.H. Lira and C.M. Vest. Refraction correction in holographic interferometry and tomography of transparent objects. *Applied Optics*, 26(18):3919–3927, 1987.
401. C.M. Vest. Interferometry of strongly refracting axisymmetric phase objects. *Applied Optics*, 14:1601–1606, 1975.
402. R. Koslover and R. McWilliams. Measurement of multidimensional ion velocity distributions by optical tomography. *Review of Scientific Instruments*, 57(10):2441–2448, 1986.
403. H. Schardin. Die schlierenverfahren und ihre anwendungen. *Ergebnisse der Exackten Naturwissenschaften*, 20:303–439, 1942.
404. B.H. Timmerman. *Holographic interferometric tomography for cmpresssible flows*. PhD thesis, Delft University of Technology, 1997.
405. M. Sakami. *Application des methodes inverses a la metrologie thermique par spectrometrie infrarouge*. PhD thesis, University de Poitiers, 1994.
406. G.N. Blinkov, D.E. Vitkin, A.F. Grishan, N.A. Fomin, and A.V. Chaley. Speckle diagnostics of gas flows from nozzles. *Izvestia Sibirskogo Otdelenia Academii Nayk SSSR*, 4:102–103, 1990.

407. I. Lira and R. Keller. Reconstruction of an axisymmetric refractive index distribution using speckle photography. In *SPIE Proc.*, Volume 2340, pages 493–502, 1994, SPIE Press.
408. I. Lira. Reconstruction of an axisymmetric refractive index distribution with non-negligible refraction. *Measurement Science and Technology*, 5:226–232, 1994.
409. T.E. Walsh and K.D. Kihm. Tomographic deconvolution of laser speckle photography for axisymmetric flame temperature measurement. *Journal of Flow Visualiztion and Image Processing*, 2:299–310, 1995.
410. A.H. Andersen and A.C. Kak. Digital ray tracing in two-dimensional refractive fields. *Journal of the Acoustic Society of America*, 72:1593–1606, 1982.
411. A.C. Kak. *Tomographic Imaging with Difracting and Non-Difracting Sources.* Prentice-Hall, Englewood Cliffs, NJ, 1985.
412. A.H. Andersen and A.C. Kak. Digital ray tracing in two-dimensional refractive fields. *Journal of the Acoustic Society of America*, 72:1593–1606, 1984.
413. R. Gordon, R. Bender, and G.T. Herman. Algebraic reconstruction techniques (ART) for three-dimensional electron microscopy and x-ray photography. *Journal of Theoretical Biology*, 29:471–481, 1970.
414. D.D. Verhoeven. Multiplicative algebraic computed tomographic algorithms for the reconstruction of multidirectional interferometric data. *Optical Engineering*, 32(2):410–419, 1993.
415. G. Minerbo. Ment: a maximum entropy algorithm for reconstructing a source from projection data. *Computer Graphics and Image Processing*, 10:48–68, 1979.
416. L. Hesselink. In W.-J. Yang editor, *Handbook of Flow Visualization*, chapter Optical tomography, pages 307–329. Hemisphere Publ. Corp., Washington DC, 1989.
417. R.N. Bracewell. Strip integration in radio astronomy. *Australian Journal of Physics*, 9:198–217, 1956.
418. R.N. Bracewell and A. Riddle. Inversion of fan-beam scans in radio astronomy. *Astrophysical Journal* , 150:427–434, 1967.
419. G.N. Ramachandran and A.V. Lakshminarayanan. Three dimensional reconstruction from radiographs and electron micrographs: application of convolution instead of Fourier transform. *Proceeding of the National Academy of Sciences of the USA*, 68:2236–2240, 1971.
420. L.A. Shepp and B.F. Logan. The Fourier reconstruction of a head section. *IEEE Trans. Nucl. Sci.*, pages 21–43, 1974.
421. S. Sato, S.J. Norton, M.J. Linzer, *et.al.* Tomographic image reconstruction from limited projections using iterative revisions in image and transform spaces. *Applied Optics*, 20:395–399, 1981.
422. S. Sato, K.Saski, Y. Nakamura, M. Linder, and S. Norton. Tomographic image reconstruction from limited projections using coherent optical feedback. *Applied Optics*, 20:3073–3076, 1981.
423. M. Freeman and S. Katz. Determination of a radiance distribution of an optically thin radiating medium. *Journal of the Optical Society of America*, 53(10):1172–1179, 1963.
424. V.V. Pickalov and T.S. Melnikova. Tomographic measurements of plasma temperature fields . *Beitr. Plasmaphysik.*, 24(4):417–430, 1984.
425. D.D. Verhoeven. Limited-data computed tomography algorithms for the physical sciences. *Applied Optics*, 32(20):3736–3754, 1993.
426. V.V. Pickalov and N.G. Preobrazhensky. Restoration of local equilibrium temperature fields over radiation. *Journal of Ingineering Physics*, 33(6):1042–1046, 1977.

427. V.V. Pickalov and N.G. Preobrazhensky. Restoration of two-dimensional radiation fields for an optically thick plasma. *Journal de Physique.*, 40(7):855–856, 1979.
428. V.V. Pickalov and N.G. Preobrazhensky. Calculative tomography and physical experiment (in Russian). *Uspechi Fisicheskich Nauk*, 141(3):469–498, 1983.
429. N.A. Fomin. Speckle tomography of gasdynamic flow. In K. Takajama, editor, *Shock Waves. Proc. of International Shock Wave Symp.*, Volume 2, pages 1019–1024, Sendai, 1992. Springer Verlag.
430. N.A. Fomin and D. Vitkin. Speckle tomography of a flame in multi-jet combustors. In J.P.Crowder, editor, *Flow Visualization VII. Proc. of the Seventh International Simposium on Flow Visualization*, pages 910–914, New-York, 1995. Begell House, Inc.
431. S. Bahl and J.A. Liburdy. Three-dimensional image reconstruction using interferometric data from a limited field of view with noise. *Applied Optics*, 30(29):4218–4226, 1991.
432. H. Hertz. Experimental determination of 2-d flame temperature fields by interferometric tomography. *Optics Communications*, 54:131–136, 1985.
433. D. Vukicevic, H. Jäger, T. Neger, H. Philipp, and J. Woisetschläger. Tomographic reconstruction of the temperature distribution in a convective heat flow using multidirectional holographic interferometry. *Applied Optics*, 28:1508–1516, 1989.
434. X. Xudong, S. Dufang, C. Shaohua, and H. Caiyuan. Measurement of temperature distribution in heat flow using holographic tomography technique and image processing system. *Journal of Flow Visualization and Image Processing*, 2(3):311–319, 1995.
435. H. Uchiyama, M. Nakajima, and S. Yuta. Measurement of flame temperature distribution by ir emission computed tomography. *Applied Optics*, 24(23):4111–4116, 1985.
436. G.W. Faris and R.L. Byer. Quantitative optical tomographic imaging of a supersonic jet. *Optics Letters*, 11(7):413–415, 1986.
437. E. Bar-Ziv, S. Sgulin, O. Kafri, and E. Keren. Temperature mapping in flames by Moiré deflectometry. *Applied Optics*, 22:698–705, 1983.
438. G.W. Faris and R.L. Byer. Beam-deflection optical tomography. *Optics Letters*, 12(2):72–74, 1987.
439. G.W. Faris and R.L. Byer. Beam-deflection optical tomography of a flame. *Optics Letters*, 12(3):155–157, 1987.
440. G.W. Faris and R.L. Byer. Three-dimensional beam-deflection tomography of a supersonic jet. *Applied Optics*, 27(24):5202–5212, 1988.
441. A. Asseban. *Caracterisation de transferts thermiques en convection naturelle, par la technique de speckle photographique.* PhD thesis, Universite de Poitiers, 1996.
442. F. Mayinger and W. Pankin. Holographic two-wavelength interferometry of combined heat and mass transfer. In R.Goulard, editor, *Combustion Measurements*, pages 270–283. Hemisphere Publ.Corp., Washington, DC, 1976.

Index

Printing: Mercedesdruck, Berlin
Binding: Buchbinderei Lüderitz & Bauer, Berlin